The History of Science
Volume 4
The European Renaissance

Dr. Peter Whitfield

First published in the United States in 2003 by Grolier Educational, a division of Scholastic Library Publishing, Sherman Turnpike, Danbury, CT 06816

For Compendium Publishing

Contributors: Sandra Forty

Editor: Felicity Glen

Picture research: Peter Whitfield and Simon Forty

Design: Frank Ainscough/Compendium Design

Artwork: Mark Franklin/Flatt Art

Reproduced by: P.T. Repro Multi Warna, Indonesia.

Printed in China by: Printworks Int. Ltd

Library of Congress Cataloging-in-Publication Data

Whitfield, Peter, Dr.

History of science / Peter Whitfield

p. cm.

Includes index.

Contents: v. 1 Science in ancient civilizations – v. 2 Islamic and western medieval science – v. 3 Traditions of science outside Europe – v. 4 The European Renaissance – v. 5 The Scientific Revolution – v. 6 The eighteenth century – v. 7 Physical Science in the nineteenth century – v. 8 Biology and Geology in the nineteenth century – v. 9 Atoms and galaxies : modern physical science – v. 10 Twentieth-century life sciences.

ISBN 0-7172-5729-0 (act : alk. paper) – ISBN 0-7172-5703-7 (v. 1 : alk paper) – ISBN 0-7172-5704-5 (v. 2 : alk paper) – ISBN 0-7172-5705-3 (v. 3 : alk paper) – ISBN 0-7172-5706-1 (v. 4 : alk paper) – ISBN 0-7172-5707-X (v. 5 : alk paper) – ISBN 0-7172-5708-8 (v. 6 : alk paper) – ISBN 0-7172-5709-6 (v. 7 : alk paper) – ISBN 0-7172-5710-X (v. 8 : alk paper) – ISBN 0-7172-5711-8 (v. 9 : alk paper) – ISBN 0-7172-5712-6 (v. 10 : alk paper)

1. Science–History–Juvenile literature. [1. Science–History.] I. Grolier Educational (Firm) II. Title.

Q125 .W586 2003

509—dc21

2002029844

Acknowledgments

The publishers would like to thank the following for their help with the illustrations: Venita Paul and Sarah Sykes at the Science & Society Picture Library, Science Museum Exhibition Road, London SW7 2DD; and Dawn Hathaway and Hillary Smith at the Natural History Museum (**NHM**—nhmp@nhm.ac.uk 0044 (0)2079425401).

Picture credits

All maps and artwork are by Mark Franklin/Flatt Art. All photographs were supplied by the Science & Society Picture Library except those on the following pages (T=Top; C=Center; B=Below): Author's collection p.4–5, p.11 (B), p.12, p.13 (British Library), p.19 (B), p.22, p.30–33, p.41, p.42, p.44, p.45, p.46, p.47 (T), p.48 (British Library), p.50, p.52, p.55, p.56, p.59, p.64; Sheila Terry/Science Photo Library p.11 (T); British Library p.26, p.43; NHM p.39.

Note

Underlined words in the text of this volume and other volumes in the set are explained in the Glossary on page 70.

Contents

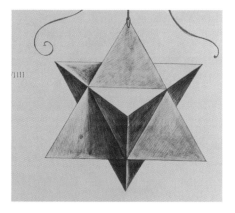

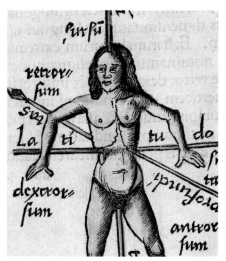

The Renaissance:
An Age of Transition

The years 1450 to 1550 witnessed a series of profound changes in European life and thought that we call the "Renaissance," a word that means rebirth. Historians have never been able to agree precisely why these changes occurred at this time, or how they affected each other. But taken together, they mark the end of the Middle Ages and the emergence of the modern world.

What were these changes? First, there was the enrichment of urban life following the devastating plagues and famines of the fourteenth century. Trade improved throughout Europe, and the new wealth was spent on private buildings and public spaces to create a new ideal of civilized living. This occurred first in Italy, but it soon spread to Germany, France, and England. The cities of the Renaissance became the fountainheads of new ideas.

Second, there was the invention of printing, which revolutionized the speed with which ideas and information could be spread. A text printed in Germany, for example, could be read within a matter of weeks by thousands of people far away in France, Italy, or England, and the printed page cost a fraction of a hand-copied manuscript. Printing was the most important change in human communication since the invention of writing itself, around 5,000 years earlier.

The system of perspective that developed during the Renaissance provided a norm of pictorial representation that holds good today.

Third, there was the discovery of the "New World" across the Atlantic, something that revolutionized people's understanding of their world overnight. This event was doubly significant because the New World had been completely unknown to all the great Greek and Roman philosophers and thinkers of the past, and it seemed to symbolize the position of "Renaissance man," standing at the opening of a new age.

Fourth, the Reformation within the church broke up forever the unity of medieval Christian Europe. The thousand-year-old power of the Roman church was challenged. The conscience and beliefs of the individual were set up as a new ideal, opposing the authority of any lawgiving institution. The role of

ʾERRÆ·MOTVS

printing in the Reformation was crucial, for if Luther's arguments had not been printed and read by thousands of others, his dispute with the church might have remained a purely private quarrel.

Fifth, there was the artistic rediscovery of nature. People looked with new eyes at the real world and learned to portray the natural world and the human form with a new fidelity. They studied the laws of perspective and were able to rationalize space in their pictures so as to produce an illusion of reality. This is the aspect of the Renaissance that is most familiar—the paintings of courts or temples filled with graceful human figures framed by hillsides, or landscapes drawn with natural form and color.

All these factors came together to create a climate of innovation, whereas the Middle Ages had been characterized by continuity and respect for authority. Was there also a movement of renaissance, of rebirth, in the world of science?

Strangely, perhaps, it is difficult to give a clear answer to that question. The "Scientific Revolution" did not really begin until the late sixteenth century, almost 100 years after the changes listed above were well established. But before that revolutions had taken place in many aspects of applied science—in navigation and mapmaking, mining, engineering, architecture, and medicine. All these revolutions sprang from a changed understanding of nature, and in this volume we will look at the ways in which these innovations shaped European history and prepared the way for the revolution in science.

Symbol of the Renaissance:
The Printing Revolution

**JOHANNES GUTENBERG
(1400–68)**

- Inventor of printing from movable type.
- Born Mainz, Germany.
- 1430–44 Worked in Strasbourg, probably as a goldsmith.
- Around 1439 started experimenting with printing.
- 1450 Started a partnership with Johann Fust in Mainz.
- Partnership ended in 1455. Fust completes the 42-line Bible.
- Financially ruined, Gutenberg managed to set up another press. Among his books he printed a school Latin grammar by Aelius Donats.

The printing press can well stand as the symbol of the Renaissance, just as the steam engine symbolizes the Industrial Revolution. As the steam engine was a new source of physical power, so the printed book was a new source intellectual energy, a way of storing and spreading knowledge and ideas throughout the world. The invention involved a solution to complicated technical problems, but behind it lay a commitment to literacy as an ideal of civilization.

The first full printed book was the Bible of 1455 issued by Johannes Gutenberg, the father of printing. Gutenberg (around 1400–68) was a native of Mainz in Rhineland Germany, a skilled metalworker, perhaps a goldsmith, who worked on his invention in close secrecy for some 15 years.

The idea of printing was not new; it had been practiced in China for several centuries before Gutenberg, and even in Europe it has a protohistory. The principle of printing was even older of course, as the use of seals and wax show, while Mesopotamian peoples used carved stone seals to print in wet clay 5,000 years ago. In medieval China wooden blocks were incised with pictures and words, and used for printing on fabric or paper. Since an entire page was cut and printed as a unit, it was time-consuming and suitable only for brief subjects—prayers, short poems, or decorative pictures were reproduced in this way. They could be printed in black ink, and color could be added afterward. It was a novelty rather than a process that was important to society.

This form of printing was tried in Europe in the early fifteenth century, in the decades immediately before Gutenberg. In the

Netherlands and Germany experiments were made using carved wooden blocks and, more interestingly, metal plates. Hard metal punches were cut and used to stamp individual letters into a soft metal such as lead, and in this way a whole page could be laid out. In practice, it was impossible to strike the letters in an even line and with an even depth, so that the printed result looked very poor—far worse than a handwritten text, and the labor involved in punching an entire page was enormous. The woodblock technique, however, was used with some success for subjects such as playing cards, although it, too, was unsuitable if there were more than a few words of text.

The printer's workshop: A technological change became the agent of intellectual revolution.

ALPHABETICAL ORDER

The originality of Gutenberg's approach lay in the idea of separating out all the elements that appear on a printed page, and combining them as necessary. In this the Latin alphabet has the distinct advantage of having only 26 letters: Out of just 26 elements, any text, no matter how long, could be constructed. So Gutenberg's fundamental problem was to produce multiple master shapes of these letters that could be made up into a text, printed, then broken up and made up into another text; this is what came to be termed "movable type."

His solution finally involved three stages. First, 26 punches were cut by hand in a hard metal such as brass; this then was used to stamp the letter into a piece of softer metal called a matrix. The matrix then formed a mold into which molten lead was poured, forming the finished piece of type. Each matrix mold could be used over and over again to produce hundreds of letters all of which were identical to each other.

The tolerances involved were tiny, since the slightest variation

Johannes Gutenberg, the German goldsmith who invented printing.

Page from an early printed book on astronomy; the picture has been cut on a woodblock and inserted into the type.

in the size of type would produce a printed page that was untidy or difficult to read. This applied, too, to all punctuation marks and numerals, while careful consideration had to be given to spacing and proportioning the letters, so that a small letter such as "i" was in proportion to "w," and so on. Gutenberg's skills as a metalworker must have been taxed to the limit in cutting the originals and reproducing them from the matrix molds.

Even with the type made, Gutenberg still faced many difficulties. He had to develop the technique of "locking up" the thousands of tiny pieces of type—that is, holding them rigid in a frame while printing. This was achieved by using screws to tighten the sides of the printing plate into place. He had to modify the winepress, something that would have been very familiar to him in the Rhineland, so that the type would bite evenly into the paper. He had to make paper that would show a clear impression without the ink spreading and becoming a blur. He also had to develop a stable ink. He did this from soot and linseed oil. The ink was so good that it remained the traditional printer's ink until the early twentieth century.

None of these elements had been addressed before in Europe or in China. In fact, the structure of the Chinese written language, with its thousands of characters, made it far less suited to movable type than the Latin alphabet. The extent to which Gutenberg solved all these problems may be seen in the clear and powerful pages of his 1455 Bible, some copies of which were adorned with marginal decorations or coloring.

Sadly, Gutenberg does not seem to have reaped any enormous rewards for his labor and originality. He quarreled with his business partner, Johann Fust, who then took over the printing press, and Gutenberg seems to have retired with a small pension from the Archbishop of Mainz.

However, his invention revolutionized the production of books, and within a very few years presses were established in other cities in Germany and in Italy, and then throughout Europe. Pictures and diagrams could be added to texts by the use first of woodblocks, then later of engraved copper plates. It was from copper plates that the first comprehensive series of world maps was printed in Italy in Bologna and Rome in 1477 and 1478. Works on astronomy, mathematics, and medicine also appeared in the 1470s.

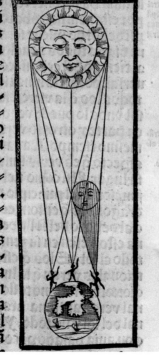

THE SPREAD OF KNOWLEDGE

The invention of the new medium of printing did not in itself produce new ideas; in fact, the earliest printed books were religious works and Latin literature. But it made classical texts in the sciences available far more widely than ever before. They were relatively cheap, and they were easily replaceable. It became far easier to compare one author with another and to compare what written authorities said with what could be seen in nature. Scholars could be sure that their text of Ptolemy—or Aristotle or Galen— was exactly the same as another scholar's in another city or another country. Knowledge and ideas accumulated, developing and progressing from one text and one authority to the next.

It has been estimated that by 1500, within fewer than 50 years of printing, some nine million books had been produced in western Europe. The impact on literacy and learning was immense; and when new ideas were put forward, the new medium of printing spread them irresistibly. After 1450 the history of ideas can be traced through the publication of certain landmark texts in fields such as astronomy, medicine, geography, or mathematics. Gutenberg's may justifiably be called the first modern invention: A solution to a series of technical problems that then went on to change society.

William Caxton, who learned printing in Flanders and brought the technique to England in 1476.

WILLIAM CAXTON
(around 1422–91)
- Translator and first printer in English.
- Born in Kent, England.
- Apprenticed to a textile merchant in London 1438.
- Moved to Bruges, then in 1471 to Cologne to learn the printing trade.
- Returned to Bruges to set up his own printing press.
- 1474 First printed book—and the first book printed in English—his translation of a French romance, *Recuyell of the Historyes of Troy*.
- 1476 Returned to London and set up printing press in Westminster.
- Printed about 100 books, including many French and Latin translations, plus two editions of the *Canterbury Tales* by Chaucer and *Morte d'Arthur* by Sir Thomas Mallory.

Mathematics:
The Concept of Ordered Space

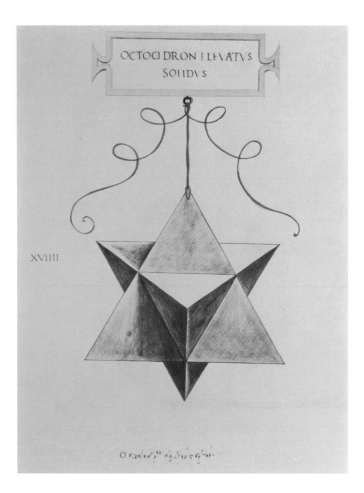

One of the geometric designs from Pacioli's *On Divine Proportion,* drawn by Leonardo da Vinci.

**LUCA PACIOLI
(around 1445–1517)**
- Franciscan friar.
- Studied theology and then became a traveling teacher of mathematics.
- 1497 Moved to Milan to work with Leonardo da Vinci.
- Still teaching mathematics, in 1494 he wrote *Summa de arithmetica, geometrica, proportioni et proportionalita,* ("Encyclopedia of arithmetic, geometry, and geometric proportion").
- 1509 Published *Divina proportione* ("On divine proportion"), a study of mathematical and artistic proportion that included some illustrations by Leonardo.

Between 1400 and 1600 mathematics was applied to a number of problems connected with physical space—new ways of measuring space and new ways of representing it in diagrams and works of art. The medium of printing spread these new techniques widely among scholars and educated laymen throughout Europe.

There was one technical change in the language of mathematics that was most important: the use of modern Arabic numerals for calculation. They were common throughout the Islamic world and had been introduced into western Europe as long ago as the thirteenth century, when Leonardo Fibonacci (around 1170–1250) wrote a book on mathematical problems. Fibonacci had learned about the numerals while traveling in North Africa with his father, an Italian merchant, and he immediately recognized their superiority over the cumbersome Latin numeral system. It took many years, however, before they were understood and used throughout Europe.

The leading characteristic of Renaissance math, however, was its concern with geometry and proportion. This concern seems to have begun with artists and architects before mathematicians were able to give it a scientific basis. The fifteenth-century painter Piero della Francesca (around 1420–92) wrote a book on perspective around the year 1470. It was never published, but it circulated in manuscript. In this work he showed how the illusion of depth could be created by tracing lines back to a vanishing point. He showed how solid shapes must be analyzed into planes, all reducing in size in direct proportion to their distance from the eye. Piero's own paintings all show human figures in architectural settings where the perspective is carefully calculated.

The ability to depict complex shapes correctly on paper became an aim in itself: Luca Pacioli (around 1445–1517) was a mathematician who wrote a work called *On Divine Proportion,* published in 1509, for which figures such as this three-dimensional image (above left) were drawn by Luca's friend Leonardo da Vinci. The author then analyzed the figures into their components of triangles and pentagons.

ART AND GEOMETRY

Both Leonardo and Albrecht Dürer (1471–1528), his great German contemporary, analyzed the human figure in terms of its ideal proportions—that the span of the arms is equal to the height, that the face from hairline to chin is one-tenth of the height, that the span of the shoulders is one-quarter of the height, and so on. Dürer also demonstrated how foreshortened images could be produced by plotting points within a coordinate system. Like Leonardo, Dürer was convinced that geometry held the key to true art. "The sole reason why painters are not aware of their errors," he wrote, "is that they have not learned geometry, without which no one can become a true artist." He also showed how, in large pictures or in architecture, allowance must be made for the observer's limited viewpoint.

Many of these artists also became interested in calligraphy and showed how the letters of a good typeface all obeyed geometric laws. Behind all these theories lies the belief that in order to represent the forms found in nature, the artist must understand them in mathematical terms.

The Renaissance concern for measurement was applied particularly to the Earth itself. The interest here was not so much in determining the globe's actual size, for there had been no

Above: Leonardo's ideal proportions.

Below: Perspective drawing of a lute by Dürer. The artist touches one end of a string on various points of the lute, and the point at which the line intersects a window is noted. The joining of these points will give a precise, foreshortened image of the lute—a projection. The window represents the eye, and the string represents individual rays of light from the lute to the eye.

theoretical advance on that problem since the Greeks (see Volume 1, page 50), but in locating places on the Earth in terms of latitude and longitude, and in placing the Earth within a celestial framework by means of poles, tropics, and equator. This science came to be known as "cosmography"—literally, the drawing of the world—and many texts entitled *Cosmographia* were published, especially in Germany and Italy. Such texts also dealt with matters such as triangulation survey, the determining of heights by means of shadow triangles, the calibration of sundials at different latitudes, and the use of all manner of quadrants, staffs, and <u>astrolabes</u> for sighting and measuring. Although some were in Latin, more and more of these texts were written in the vernacular languages as the century progressed, so as to reach people in practical professions such as surveyors, architects, and navigators. An English version called *Pantometria* (meaning "The Measuring of All Things") was published by Thomas Digges (died 1595) in 1571.

An artist draws a landscape by plotting each feature on squared paper—a coordinate system. This picture also shows the Renaissance fascination with perspective.

THE MATHEMATICS OF WAR

New technology in warfare during the Renaissance produced another area in which mathematics became of great importance: ballistics. Powerful cannons were too heavy to be maneuvered quickly in pitched battles, but they did transform the art of siege. To understand how the cannonball traveled, it was vital to know how to mount and aim the cannon.

The classical physics of Aristotle stated that a projectile traced two sides of a triangle: It ascended in a straight line until its force was exhausted, then it simply fell to earth in another straight line. But when the path of a cannonball was observed, this was clearly seen to be wrong. Some mathematicians, such as Nicolo Tartaglia (around 1500–57), clung to the triangle model, but asserted that the ball ascended in one line, described a brief arc of a circle, then descended in another straight line. From this he concluded that the maximum range for a cannon would be achieved by shooting at an angle of 45 degrees. Others suspected that the path of a projectile such as a cannonball was an arc throughout its length, although they could not analyze it precisely. Leonardo

Military engineers measure the angle of a gun barrel in order to achieve a precise trajectory; ballistics became an important science in the Renaissance.

da Vinci made sketches of the paths of cannonballs, comparing them to jets of water that moved in steep or shallow arcs depending on the angle at which they were aimed. Concepts such as force, resistance, and momentum had not yet emerged, but the mathematicians and engineers who analyzed these problems had shown that some traditional ideas in physics were mere theories that could be disproved by experiment.

PRACTICAL APPLICATIONS

Mathematics in the Renaissance developed in the context of practical activities such as draftsmanship, navigation, mapmaking, surveying, and ballistics. The many illustrated scientific texts of this period all seem to show a delight in human beings' power to master the surface of their world by measurement and calculation. Just as the artists of the Renaissance were producing magnificent portrayals of natural forms, recreating the world of nature in two dimensions, so mathematicians and mapmakers were showing how space could be ordered and analyzed by means of geometric rules. Nature was conceived to be ruled by proportion, regularity, and balance. The mathematical description of force and movement was more difficult and still lay in the future; but by looking at nature with fresh eyes, the artists and scientists of the Renaissance were steadily extending the language of mathematics.

ALBRECHT DÜRER (1471–1528)
- Illustrator and painter.
- Born in Nuremberg, Germany.
- Trained as a draftsman, then apprenticed to Michael Wolgemut, a local painter and woodcut illustrator.
- 1490 Completed earliest-known painting, a portrait of his father.
- 1490–94 Traveled to the Netherlands, France, and Switzerland.
- 1494 First of three visits to Italy.
- 1497 Set up a studio in Nuremberg.
- 1498 Published his first great woodcut illustrations showing the Apocalypse; followed them with copperplate religious studies.
- 1512–19 Worked mainly for Emperor Maximilian I—43 pen and ink drawings for Maximilian's prayer book are among his great works.
- By 1515 had achieved international recognition and exchanged works with Raphael.
- Appointed court painter to Charles V.
- Became a devoted follower of Luther.

The Artist as Scientist:
Leonardo da Vinci

"The eye, the chief window of the soul, is the chief means whereby the understanding can most fully appreciate the infinite works of nature."

This statement by Leonardo da Vinci sums up the new approach to nature that was characteristic of the Renaissance, and that linked the artist to the philosopher of nature. Leonardo believed that only by studying the human form, plants, animals, and physical forces with an open mind, rather than accepting the dogmas inherited from an earlier age, could nature be understood. This understanding was to be used by the artist when depicting natural forms, but the artistic motive was only a starting point in a deeper philosophical study of nature.

In a sense, Leonardo (1452–1519) has no formal place in the history of science, for he published no books on the subject, had neither teachers nor pupils, and had no direct influence on later scientists. Yet the notebooks that he kept for years are full of his

Various wheels for raising water, from Leonardo's notebooks.

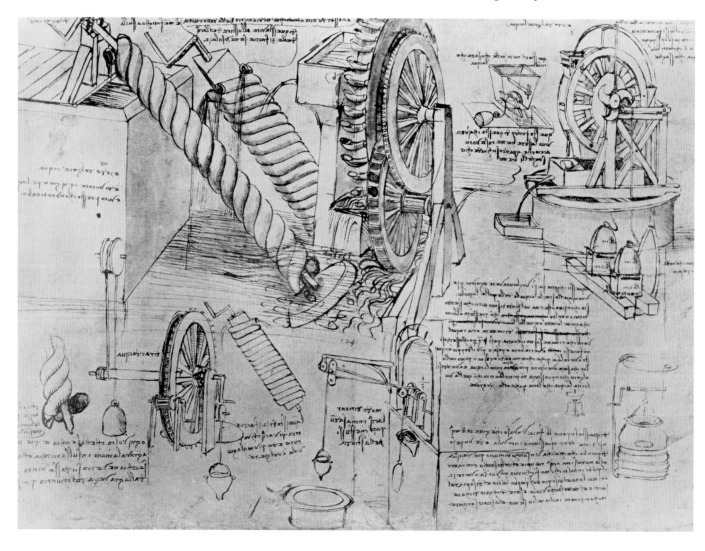

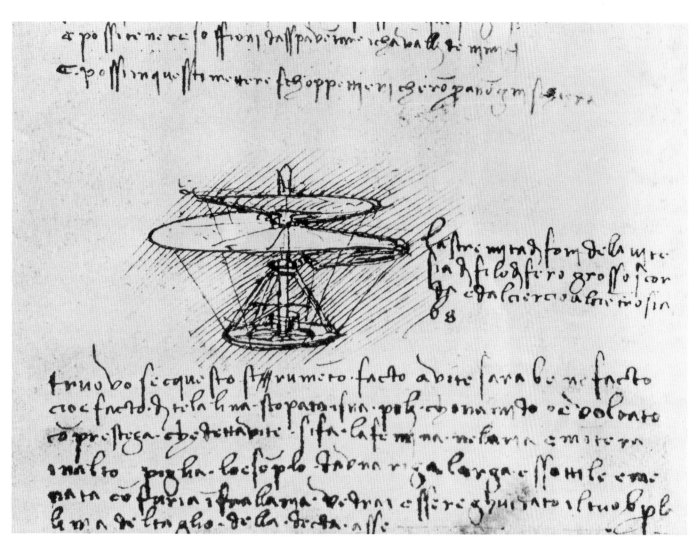

private researches into the life sciences, physics, Earth sciences, and technological ideas. These manuscripts, which only became generally known in the nineteenth century, reveal him to be the archetypal mind of his age, and they reveal the extent to which new ideas were welling up in the Renaissance.

Did Leonardo envisage the helicopter? This drawing suggests that he did.

UNDERLYING PATTERNS

Leonardo's overwhelming concern was for the mathematical laws that he believed underlay all nature. His manuscripts are full of drawings and notes about a multitude of different subjects, all apparently mixed together on the page. This was because he would study any topic by amassing as many diverse examples of it as he could. Thus a page devoted to curves might include sketches and notes on waves, grasses, hair, sails, or screw threads. He studied these things in order to be able to paint them, but also to find their underlying patterns and unity.

"Painting compels the mind of the painter to transform itself into the very mind of nature," he wrote. He believed mathematics— especially geometry—to be the hidden laws of nature. He was not a skilled mathematician, but he believed that "there is no certainty of knowledge where one cannot apply any of the mathematical sciences,"

Leonardo as a young man.

Leonardo da Vinci, whose art sprang from a scientific vision of nature.

LEONARDO DA VINCI (1452–1519)
- Renaissance genius.
- Born Vinci, Italy.
- Around 1470 became a pupil at the studio of Andrea del Verrocchio. While there painted *Baptism of Christ* and *Adoration of the Magi.*
- 1482 Moved to Milan to work for Duke Ludovico Sforza, painting and directing court pagents. Devised a hydraulic irrigation system to bring water to the Lombardy plains.
- 1498 Painted the *Last Supper.*
- When Duke Ludovico lost power, moved to Florence, 1500. Entered service of the duke of Romagna, Cesare Borgia, as his architect and engineer.
- With Michelangelo was commissioned to decorate the Sala del Consiglio in the Palazzo della Signoria.
- Completed the *Mona Lisa* in about 1504.
- 1506 Worked for Louis XII of France.
- Few of his paintings have survived, but he left many drawings and sketches. Filled numerous notebooks with sketches and plans and copious notes written in mirror writing.
- 1516 Francis I granted him an annual pension and let him use the Château Cloux for life.

and that "proportion is found not only in number and measurements, but also in sounds, weights, times, and spaces." His notebooks contain many geometric representations of motion, weight, and speed.

Not surprisingly he was especially drawn to the study of vision, and he suggested that light was reflected from any object in circles that spread like ripples in still water. "Just as a stone flung into water becomes the center and cause of many circles," he wrote, "so any object placed in the luminous atmosphere diffuses itself in circles and fills the surrounding air with images of itself." His interest in vision led to him to dissect the eye and the human brain, and he speculated if the mind's many functions—memory, imagination, sight, conscience, and so on—had their own specific locations in the brain.

UNDER THE SKIN

Leonardo's attention to anatomy left its clearest mark on his art, which showed an unprecedented mastery of human form. His devotion to truth and nature in this field cost him much anguish, for he records "passing the night hours in the company of corpses, quartered and flayed, horrible to behold." His dissections led him to theorize on the body's physiology, and he made models of the heart from wax and from glass in an unsuccessful attempt to understand its workings.

Equally original were his investigations into the mystery of flight: He made detailed studies of the bones and muscles of birds, and tried to extract a formula to describe the relationship between weight and wingspan that he might then apply to humans. Experiment was crucial to Leonardo, and it was another demonstration that he was ahead of his time. He would place a man with artificial wings on a huge set of scales in order to measure how much weight was displaced when the wings were flapped. He would place markers such as seeds or dyes in water in order to observe the patterns of a current. If an experiment falsified an accepted rule, he would reject the rule: For example, he disproved experimentally the Aristotelian rule that "if a power moves an object with a certain speed, it will move half that object with double that speed."

DESIGNS FOR LIVING

In his own time Leonardo was known as a professional engineer and surveyor, rather than as a painter. His inventions were numerous, although most of them were mere ideas or designs and could not actually be built. He designed gears, bearings, lathes, presses, canal locks, textile-spinning machines, thread cutters, weapons, and flying machines. He made a form of plastic by impregnating layers of paper with gelatin and egg white, producing hard, water-resistant, unbreakable plates. His grasp of mechanics was excellent, and he knew that the quest for a perpetual motion machine was an illusion, because no machine generates its own energy out of nothing.

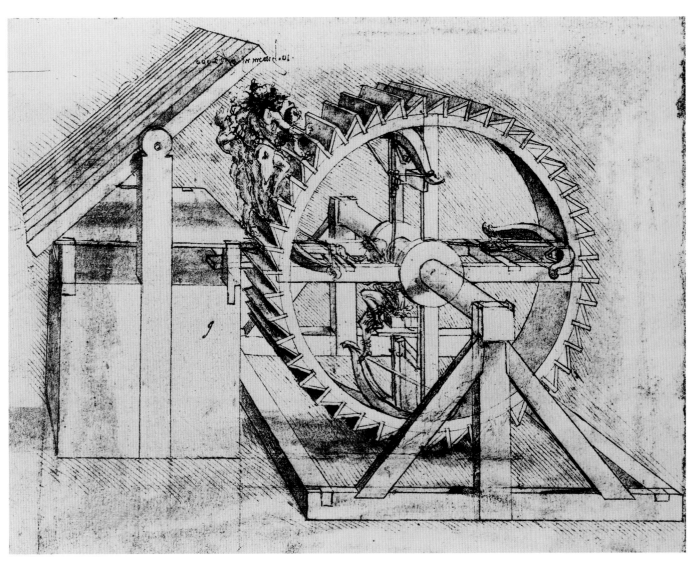

Design for a gigantic mechanical crossbow, loaded by the power of a treadmill.

Leonardo does not seem to have speculated a great deal about the heavens or the structure of the cosmos, but he was deeply interested in geology. He drew an elaborate parallel between humankind and the Earth: "As man contains within himself bones that support the flesh, so the world has rocks that support the soil; as man has in him the lake of blood that the lungs swell and decrease in breathing, so the body of the Earth has its oceanic sea that also swells and diminishes every six hours in nourishing the world."

He made a model of the Mediterranean Sea in order to study the effects of the rivers that flow into it. He reasoned that the face of the Earth had been formed by the action of water, wind, frost, and sedimentation over thousands of years, and had not been formed in an instant as in the biblical account. Although most of his ideas remained as notes and sketches, Leonardo was a prophetic figure whose creed was observation and experiment in the search for the laws and logic in nature's structure.

Leonardo's German contemporary Albrecht Dürer was equally convinced that law and proportion were the keys to nature's structure, and he sketched many experiments showing the mathematical laws of perspective and the ideal proportions to be found in nature.

Valturio and the Science of War: *De re militari*

Opposite, Above: A primitive tank designed by Valturio; the wheels were turned by a worm-driven axel, but the crank-turners at the rear must have been protected.

Opposite, Below: A siege rampart lowered by means of a toothed wheel.

Below: A fantastic siege-engine designed by Valturio to terrify the enemy. Soldiers inside fired cannon, then poured out of the drawbridge. It was winched forward by pulleys.

Technology and fantasy came together in a landmark work of Renaissance book publishing, Roberto Valturio's *De re militari*—"On Military Matters," printed in 1472. It is a textbook on the techniques of warfare written by a scholar who was a courtier to the duke of Rimini in Italy. As far as we know, Valturio himself (1405–75) was not a soldier, and he wrote the book from a purely historical and theoretical point of view. Its main interest lies in the dozens of ingenious machines that it describes, illustrating them with delightful woodcuts. Some of them, perhaps most of them, can never have been built: They were just ideas, but they show a mind experimenting with physical forces, designing new solutions to old problems.

Among the unusual offensive weapons that Valturio presents are a mechanical spear launcher in which a long length of wood acts like a great spring to hurl the spear forward, and a form of tank, an armor-plated vehicle with a battering-ram nose, that seems to suffer from the defect that its wheels are turned by cranks on the *outside*.

Many of the designs show forms of siege engines—that is, means of scaling the walls of an enemy town. One such device takes the form of a wooden tower that can be raised or lowered by levers in a parallelogram action. The most eye-catching of these machines is a huge metal-plated dragon that moves on rollers; it carries a cannon in its mouth, and a drawbridge can be lowered from its belly. Presumably its appearance was designed to strike terror into the enemy. Valturio has a lengthy section on launching attacks by water, including a collapsible and portable boat built in sections. Most remarkably, there is a submarine worked by paddles from the inside. Many of these

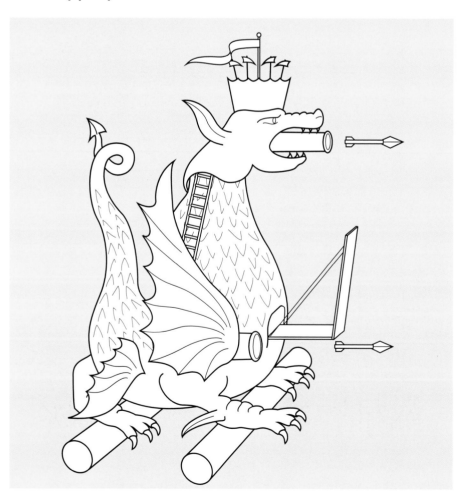

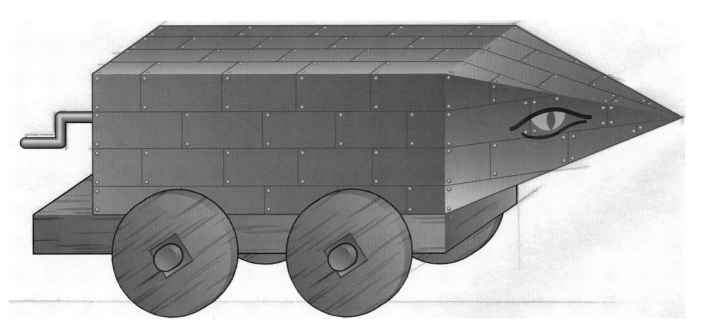

machines employed cog wheels, and interlocking drives—for example, the worm-thread that turns the drive through ninety degrees.

MEN AND MACHINES

Valturio's book was widely circulated and studied by princes and army commanders, although whether they took it seriously and tried to build his machines seems open to doubt. We know that Leonardo da Vinci owned a copy, and it seems to have inspired some of his ingenious designs. Valturio's weapons may have existed only in his mind, but they show a new confidence in people's ability to augment their natural strength by means of machines. The practical problems of construction are not dealt with, and there is no theoretical analysis of power or weight. In these fantasy designs Valturio seems to be asking the question "Why not?" As engineers attempted to create such devices in reality—for example, the pumping, lifting, and levering devices used in mining—they would discover why not, what worked effectively and what did not, and hence a new language of forces would be developed.

Gems of Wisdom: A Renaissance Encyclopedia of Science

A surveyor using a Jacob's staff to measure distances.

Scientific thinking was joined to the new art of printing in a delightful book published in 1503 entitled *Margarita Philosophica.* This means literally "The Philosophic Pearl"—perhaps we should call it colloquially something like "Gems of Wisdom." Its author was Gregor Reisch (around 1467–1527), a teacher at the university of Freiburg in southern Germany. Reisch's aim in this book was to bring together current ideas about the way the natural world functioned, and the result was an encyclopedia of science on the eve of the Renaissance. The subjects covered by Reisch were:

- The divisions of learning—theology, mathematics, grammar, and so on.
- The creation of the world by God and its division into Earth, heaven, and hell.
- The human body and how it functioned.
- The shape and extent of the Earth, with maps.
- Geographical phenomena such as rainbows, earthquakes, and tides.
- The realm of nature—plants, animals, and minerals.
- Mathematics, including techniques of measurement.
- Astronomy—the structure of the celestial spheres.

This systematic description of the world evidently grew out of the medieval practice of writing commentaries on the first chapters of Genesis. With the exception of the structure of learning and mathematics, Reisch's contents strongly resemble the six days of creation, when God created in turn earth, sea, heaven, animals, plants, and finally humans. Reisch has given a scientific form to this traditional account of the universe. His explanations are not particularly original, but they sum up clearly the knowledge of his time.

THE EARTH AS A SPHERE

It is evident from Reisch's diagrams of solar and lunar eclipses that the Earth was known to be a sphere. He explains the eclipse of the Sun by correctly positioning the three relevant bodies in the order Earth–Moon–Sun, while the order for the lunar eclipse is Sun–Earth–Moon. In these diagrams the Earth is considered to be the center of the cosmos. He gives a simple <u>empirical</u> proof of the Earth's sphericity: The lookout, high up on a ship's mast, will see land features before they are seen by a sailor on deck. This had been noticed and understood by ancient Greek thinkers such as Aristotle.

In his section on mathematics Reisch describes a familiar technique for estimating the height of buildings or trees using an instrument called a Jacob's staff (see illustration at left). It was a ruler equipped with a vertical sighting rod. The sighting rod was moved until it covered the object to be measured when it was furthest from the observer's eye. The observer then moved forward a certain distance and repeated the procedure with the sighting rod; the distance between the two sighting positions on the ruler was found to be the height of the object.

Also in the mathematical section, Reisch includes a gruesome-looking figure of a man transfixed by three spears (see illustration at right). This is not a medical text on treating battlefield wounds, but an illustration of the three dimensions of any solid figure: height, width, and depth. Reisch's commentary extends this into the concept of a coordinate system in which values can be specified in three dimensions.

Reisch's encyclopedia was reprinted many times during the next 50 years, for it provided a convenient summary of traditional ideas before the revolution in science.

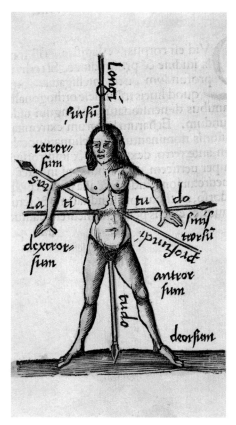

Above: This man is not the victim of a gruesome execution, but rather he illustrates the use of coordinates in three dimensions.

Left: The personification of mathematics presiding over astronomers and surveyors.

New Sciences:
Navigation

Of the new sciences that emerged in the Renaissance, none had greater practical and historical consequences than navigation. Between 1450 and 1550 European seafarers crossed the Atlantic, Pacific, and Indian oceans, and made contact with civilizations in America, India, and China. Where separate cultures had existed for centuries, an interconnected world began to develop. The European powers laid the foundations of the trading empires that were to flourish—with both good and evil consequences—for four centuries. None of this would have been possible without the navigational techniques discovered in the late fifteenth century.

THE SCIENCE OF NAVIGATION

Sailors had found the means to navigate across open sea before this: The Polynesians had colonized the Pacific islands (see Volume 3, page 18) and the Vikings had crossed the Atlantic; but exactly how they did this has never been established. European traders of the Middle Ages always preferred coastal sailing. In the Mediterranean Sea, the North Sea, or the Baltic they tried to stay within sight of coastal landmarks that were familiar to them. They could gauge the fundamentals of direction by the position of the Sun by day and the Pole Star by night. But on the open sea in cloudy weather they could only proceed by "dead reckoning"— that is, by recording how far they had traveled from their starting point in a given direction, and so estimating where they were. All this was a type of craft skill, learned by example; it did not stem from a scientific understanding of the shape of the Earth or from theoretical principles.

What was new in the fifteenth century was the application of science and mathematics to the problem of position finding. The key to this new approach was the concept of the Earth as a sphere

Magellan entering the Pacific; surrounded by demigods and sea-creatures, he calmly studies his navigational instruments

lying within the sphere of the heavens. Since the thirteenth century university scholars had been teaching spherical astronomy (see Volume 2, page 44), and all medieval scientists knew that the Earth was a sphere. The "flat-Earth" theory is a modern myth that probably arose from a misunderstanding of medieval world maps such as the Hereford *Mappa Mundi* (see Volume 2, page 66). But the world of the university scholar was quite distinct from that of the seafarer, and the idea of using science to steer a ship emerged only in the late fifteenth century.

FIXING POSITIONS

Two important innovations acted as preconditions for this change. One was the invention of the magnetic compass in the fourteenth century. This made it possible to keep a ship on a constant bearing. The other was the rediscovery around the year 1450 of the geographical theories of Ptolemy. This revived the notion of <u>latitude</u> and <u>longitude</u>, which had been forgotten in Europe throughout the Middle Ages. When seafarers were shown the idea of the globe of the Earth, measured into degrees of latitude and longitude and lying within the sphere of the heavens, they came to understand that this model could be used to fix their position at sea. The positions of stars on the celestial sphere would appear differently in relation to the horizon depending on where on the Earth the observer was standing.

The theory was not too complicated, at least in the case of latitude. The Pole Star stood at the summit of heaven, almost directly above the North Pole of the Earth. At any point on the Earth's surface away from the North Pole the Pole Star would, therefore, appear to be lower in the sky. The formula is that the elevation of the Pole Star above the horizon is the observer's latitude. For example, somewhere in the North Atlantic at the latitude of around 45 degrees north (45°N) the Pole Star would lie at 45 degrees above the horizon.

A sighting device was needed to measure this elevation, and it appeared in the mariner's astrolabe—a heavy circle of metal marked with 360 degrees around its circumference and with a metal bar across its center with which the Pole Star could be aligned. But the Pole Star

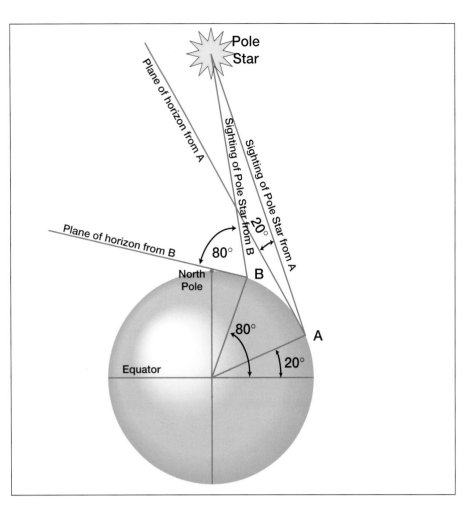

The determination of latitude by the height of the Pole Star above the horizon was fundamental: The lower the Pole Star, the further south the seafarer was.

The *Santa Maria*, Columbus's ship, was tiny by modern standards.

vanishes from sight around the equator, and there is no exact equivalent in the southern sky. The elevation of the Sun could be measured in the same way, but this was more complex because the Sun's height varies through the year—except on the equator itself—and tables had to be compiled that gave the Sun's elevation at noon at different latitudes for each day of the year.

PRACTICAL APPLICATION

Having established the principle of latitude finding, how was it used in practice? When Portuguese or Spanish seafarers first ventured out into the Atlantic from the 1490s onward, they would first estimate the latitude of the place at which they wished to land. For the Portuguese this might be the Cape of Good Hope at 35°S. Leaving European waters, they could sail south, taking daily sightings of the Pole Star or the Sun to establish their latitude. When they had arrived at 35°S, they would shape a course to the east, keeping always at that same latitude until they reached their destination. This technique was called "sailing down the latitudes," and it had the great advantage that any course could be followed to take advantage of winds and currents so long as the correct latitude was maintained. In practice, the Portuguese found that a strong southwesterly current in the mid-Atlantic would take them to the Brazilian coast before they needed to turn east, and that this was much more favorable than fighting opposing winds and currents down the western coast of Africa. Columbus, too, had reasoned that if he sailed due west from the Canaries down the latitude of 28°S, he must reach the coast of China—instead, he discovered America.

Mariners estimating distances by triangulation.

Longitude was a much tougher problem because, unlike latitude, it has no objective marker in the sky, for the celestial sphere is constantly revolving in the plane of longitude. Longitude is measured as a function of time—each 15 degrees of longitude is one hour's difference in the position of the celestial bodies. However, the absence of accurate clocks that could function at sea meant that this principle was of little practical use in the fifteenth century. For two centuries longitude could only be estimated in the same way that dead reckoning was.

THEORY AND PRACTICE

Because seafaring was a practical skill passed on from master to pupil, it took some years for theoretical works on navigation to appear in print. The foremost authority was the Portuguese Pedro Nunez (1502–78), but his works were mathematically fairly advanced, and they must have been beyond the reach of most mariners. More accessible was the work of the Spaniard Pedro de Medina (1493–1576) whose *Arte de Navigar* ("The Art of Navigation") appeared in 1545, dealing with all basic aspects of celestial navigation. It was translated into French, English, Italian, and German, and it became the model for many other texts on navigation, such as *The Seaman's Secrets* (1594) by John Davis, discoverer of Davis Strait between Greenland and Canada.

The basic astronomy described by Medina was Ptolemaic. It supposed that the Earth lay at the center of the celestial sphere and that all the heavenly bodies revolved around it. This was copied by Davis in his version, for the Copernican theory had not gained universal acceptance even by 1600.

In fact, for the purposes of observation and navigation it makes no difference whether the Earth or the Sun is the center of the universe, since the navigator is concerned solely with positions on the celestial sphere as observed from the Earth. Navigation provided a special link between theoretical and practical science, a link based on a mathematical view of the Earth.

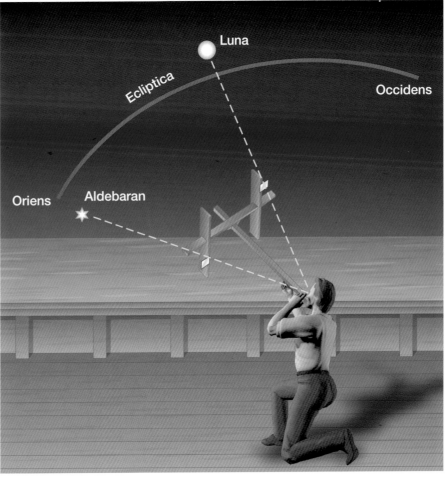

A navigator measuring the angular distance between the Moon and a bright star; such observations had to be used with nautical tables, which would fix the position.

Renaissance Geography:
Maps and the Expanding World

The world map of the medieval scholar was the *Mappa Mundi* (see Volume 2, page 66). It was geographically inexact and showed only Europe, North Africa, and western Asia; it was as much an icon as a map, showing many legends and biblical scenes. But from the fourteenth century a very different type of map began to appear in the seafaring countries of southern Europe. It was a seachart of the Mediterranean, very accurate in its depiction of coastlines and crisscrossed with lines of direction that were obviously designed to be used with the magnetic compass.

The advent of these charts is quite mysterious. We do not know how or by whom the first ones were drawn; but once the first prototypes had appeared, they could be copied, recopied, and gradually improved. They were produced in small mapmaking studios in Spain and Italy from around 1325 onward, drawn on vellum. This was like a fine water-resistant leather and was used at sea by Mediterranean seafarers. The magnetic compass appeared in Europe at the same time, but again its origin is unknown. It is thought to have been borrowed from Arab sailors, who in turn learned its use from the Chinese.

Throughout the fifteenth century many fine and detailed charts were drawn, extending north to the British Isles, east to the Red Sea, and southward beyond the Canary Islands. These charts were very

The coasts of Africa, 1502, mapped with astonishing accuracy by Portuguese mariners between 1480 and 1500.

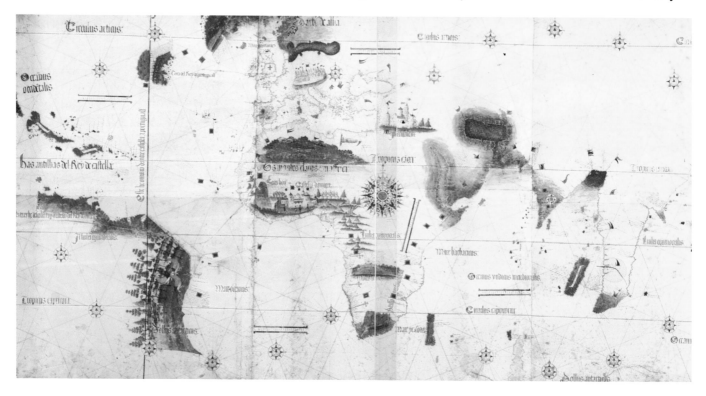

accurate for the Mediterranean coasts, but less so elsewhere. They were not scientific maps, for they had no constant scale, they did not show latitude and longitude, and they were not drawn on any projection—that is, they made no allowance for the curvature of the Earth. Designed for use by skilled pilots within the confined waters of the Mediterranean, these defects could be tolerated.

The Atlantic Ocean, a Portuguese chart of 1558; note the lines of compass bearing and the latitude bar clearly marked, but no longitude.

OCEAN GOING

By the late fifteenth century, however, Portuguese and Spanish mariners were venturing out into the Atlantic in search of new trade routes to the Far East. The seachart expanded from its Mediterranean base to take in the whole of Africa, the Atlantic Ocean, the newly discovered coasts of America, and even, a few decades later, the vast spaces of the Pacific. In other words, the seachart of Europe was transformed into a new world map. In order for this to happen, the map had to acquire a more scientific framework. It had to take account of matters of scale, measured latitude and longitude, and projection. Mapmakers were able to address these problems thanks to the rediscovery of Ptolemy's texts on mapmaking (see Volume 1, page 56) that had been unknown in the West for more than a thousand years. Scholars brought Greek texts of Ptolemy from libraries in Constantinople to Italy around the year 1450. They were translated into Latin, and the maps were reconstructed.

Ptolemy's works explained how places should be located by means of the grid of latitude and longitude, how constant proportion, that is, scale, should be applied over the whole map, and how the curvature of the Earth should be represented on a flat surface. They were, in other words, manuals of scientific mapmaking, and they became the first printed maps in Europe from the 1470s onward. Of course, Ptolemy's texts had been composed in the second century A.D., and therefore they showed only the world as it was known in the classical period—Europe, North Africa, and western Asia, as with the *Mappa Mundi*. Indeed, it was paradoxical that Ptolemy's maps should be revived and become so influential just when their view of the world was being revolutionized. Scholars in Germany and Italy set about producing world maps that would incorporate the new discoveries of the Renaissance navigators, but within the scientific framework provided by Ptolemy. These maps offered startling new images of a world that seemed to be expanding dramatically in size and richness.

PROBLEMS WITH PROJECTION

Perhaps the most difficult technical matter that mapmakers and sailors had to grapple with was projection. The Earth is a sphere, and the surface of a sphere cannot be simply spread onto a flat sheet without distortion. In charts of a small area such as the Mediterranean, covering only ten degrees of latitude, this was not so important; but when the whole of the Atlantic was being represented on a map, it became vital. The basic problem was that value of a degree of longitude decreases as one moves away from the equator. At the equator it is approximately 66 miles, at 60 degrees north it is only 32 miles, while at the poles it is zero. The value of a latitude degree, however, is constant for all practical purposes. In other words, the ratio between the value of a latitude and longitude degree varies from 1:1 at the equator to 1:0 at the poles.

The simplest way to draw any map is to divide it into squares of latitude and longitude with a constant ratio between the two, one degree by one degree, and enter locations on it. But a map drawn in this way will be totally inaccurate and fail to reflect the curvature of the Earth. The fundamental defect of the seacharts from the fourteenth to the sixteenth centuries was that no allowance at all was made for this problem. The most important result was that a straight line on one of these charts was not a straight line on the Earth, but a curve. A sailor navigating with one of these charts would have to make constant changes of course as he attempted to follow a line from his place of departure to his destination. To overcome this problem, the ratio between latitude and longitude must either be varied over the entire map, or the map must make some other compensation.

MERCATOR

This problem troubled mariners and mapmakers throughout the sixteenth century until the solution was devised by the great Flemish cartographer Gerard Mercator (1512–94). In 1569 Mercator published a large world map that had been plotted according to mathematical rules. Its aim was to allow mariners to plot a direct course from place to place by making a straight line on the map correspond to a line of constant bearing on the sea. Mercator achieved this by keeping the ratio constant between latitude and longitude, but by stretching the map as it progressed away from the equator toward the poles. The value of the longitude degree was not

Above: Compare this 1608 map of Southeast Asia with the maps on pages 26 and 27. Within a century they have become substantially more helpful to mariners. This shows the Moluccas, the spice islands that were the great goal of the European explorers.

Right: The Mercator Projection. The direct route between two points is always a great circle of the Earth, but a great circle cannot usually be shown on a map as a straight line. Mercator's map was designed so that a line of direct bearing could be drawn between any two points; this was not the shortest route, but it was by far the easiest to navigate. Mercator's map was carefully calculated mathematically for this single purpose, but it distorted the distance scale, and it was many years before mariners understood it correctly.

reduced on the map but increased, and therefore the value of the latitude degree was increased in exact proportion to keep pace with it. The visual result of this was that land masses became larger and larger the further they were from the equator; in other words, the scale increased.

The Mercator world map was neither more nor less accurate than any other world map of its time, but it had certain mathematical properties that were very important to navigators, mainly that a straight line on the map was a line of constant bearing. A captain could draw one straight line at the outset of his voyage and keep his ship headed always on that course to arrive at his destination. But the Mercator map had certain defects too. Its scale varied, so that one could not measure a linear distance on it except at the equator. It is often said that the Mercator map "shows true direction," but this is not correct. On a Mercator map a straight line is not the shortest distance between two points, and this can be seen by transferring that line onto a globe. These defects would be important if the Mercator map were being used as an image of the world, but that was not its intention. It was designed as a navigational tool, and as such it was a triumph of mathematical method. It shows the way in which Renaissance science developed by looking with new eyes at problems in practical areas previously governed by traditional craft skills.

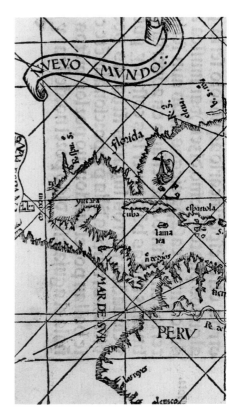

Part of a plate from a Portuguese book on navigation showing the Americas.

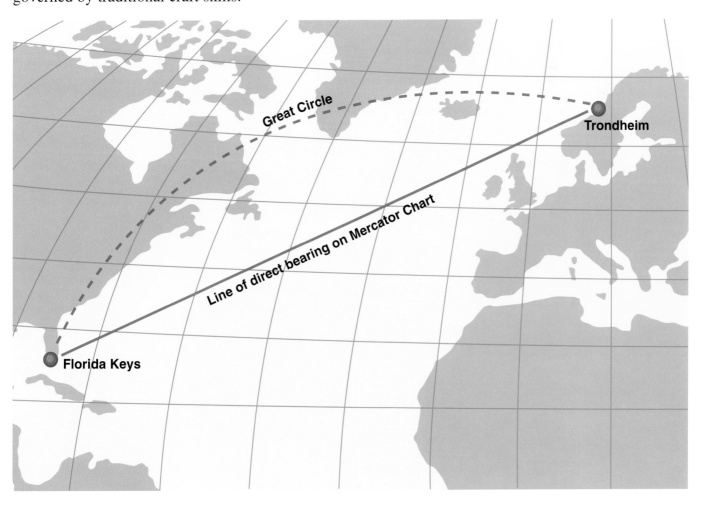

Renaissance Geography:
The Age of Discovery

The age of discovery revealed the wider world beyond Europe, but at the same time, the new medium of printing created a type of literature that had never existed before—descriptive geography. Between 1460 and 1600 a number of important texts were published in which the nations of Europe were described: Their geographical features and their history were set down for the first time in a popular, accessible form. In its early stages this literature was not especially scientific, but it created a widespread awareness of geography among a large audience, and it led in the following century to a more careful investigation of physical and human geography.

The first such book was a landmark in the history of book publishing. It was called the *Weltchronik,* or in Latin the *Liber Chronicarum.* Its author was a scholar from Nuremberg in Germany, Hartmann Schedel (1440–1514), and it is usually known as the *Nuremberg Chronicle.* This massive work ran to almost one thousand pages, and it was illustrated with hundreds of vivid woodcuts showing cities, monarchs, animals, legends, and natural wonders. It functioned like a world encyclopedia in which history and descriptive geography are blended, but its approach was still medieval in many ways.

To Hartmann Schedel the history of the world was really that found in the Bible, beginning with God's creation of the universe, continuing through the Old and New Testaments, before moving on to the modern (to us, the medieval) history of Christian Europe. The descriptions of the countries mingle fact with legend and are illustrated with delightful pictures representing the main cities. Some of the pictures, the German ones in particular, contain a degree of truth, but most of them are quite imaginary. Schedel's great book

Above and Below: Myth in an age of science: the opening up of the world's oceans did not instantly dispel the myths of the past. The illustration above is taken from the *Nuremberg Chronicle,* the one below from Munster's *Cosmographia.*

GENVA

was published in July 1493, just months after Columbus arrived back in Spain from his historic voyage to the New World; the *Nuremberg Chronicle* can thus be seen as a last view of the world when that world consisted only of medieval Europe, with its saints, its legends, its castles, and its crowded walled cities.

The city of Genoa, one of the many magnificent pictures of European cities from the *Nuremberg Chronicle*.

NEW WORLD-VIEW

Half a century later another German scholar published an encyclopedia of geography that reflected the dawn of the Renaissance. Sebastian Munster's (1489–1552) *Cosmographia* appeared in 1544, and its appeal was so great that it was reprinted around 40 times in the next century and a half. *Cosmographia* retains only small traces of the *Nuremberg Chronicle*'s structure. The biblical stories do appear—the Old Testament patriarchs are said to be the first inhabitants of the world—and there are plenty of colorful legends, but the descriptive geography is now placed within the context of a scientific understanding of the world. Munster includes maps of the four continents, of the countries of Europe, of major regions within countries, and views of cities that are now accurate enough for us to say they must have been drawn on the spot.

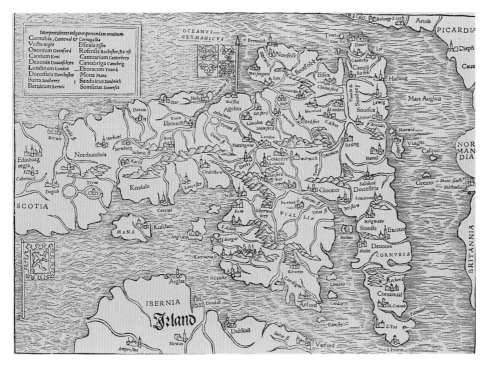

Map of England and Wales from Munster's *Cosmographia*. East is at the top of the map.

The medium-scale maps show latitude and longitude, and give a scale in German miles. General principles of geography are discussed: For example, Munster describes volcanoes, hot springs, earthquakes, and mining for metals. The sphericity of the Earth is demonstrated, and the main ideas on map projections are copied from Ptolemy. Methods of surveying and measurement, such as triangulation, are set out.

Most of the book is taken up by the countries of western Europe, but Munster includes lengthy sections on Africa, Asia, and the New World, with town plans of Mexico City and Cuzco, the Peruvian capital. Although the text still includes many traditional legends connected with the history of each nation and the natural marvels to be found there, they are now balanced by factual descriptions of geographical features, gazetteers of towns and cities, genealogies of kings and queens, and historical narrative.

LEARNING ABOUT EACH OTHER

Cosmographia set new standards in the way it linked descriptive geography with the scientific apparatus of "cosmography." As both knowledge and the sophistication of readers in Europe increased, these descriptions of the whole world were replaced by works on individual countries or regions written by scholars of the country. One of the earliest was a work on Scandinavia called *The History of the Northern Peoples,* published in 1555 by Olaus Magnus (1490–1557), who was the Catholic bishop of Uppsala in Sweden. It was the first work to describe in detail many aspects of life in the northernmost regions of Europe, such as the *aurora borealis* (also known as the northern lights), the midnight sun, the volcanoes on Iceland, the use of skis and sleds to cross the ice, the sun drying of fish, the herding of reindeer, and so on.

Earlier, in 1539, Olaus Magnus had produced a magnificent map of Scandinavia illustrated with pictures of battle scenes, enthroned kings, shipbuilders, and some wicked-looking sea monsters. These monsters on the typical maps of this period, and the many legends and magical stories still found in texts such as Munster's *Cosmographia,* raise interesting questions about such beliefs and when they died out.

MYTHS AND MONSTERS

As we saw in connection with the bestiaries and the medieval *Mappa Mundi* (see Volume 2, pages 64 and 66), a large body of such legends and beliefs had grown up in the ancient and medieval world—about strange creatures to be found in Asia or Africa. There was also a geographical equivalent to this in the shape of legendary islands or countries inhabited by half-human species. These legends could obviously flourish while a large part of the world remained unexplored, but what happened in the sixteenth century as Europeans traveled over more and more of the world? Did the belief in these legends vanish swiftly during a few years as explorers failed to find such creatures? As we can see from the maps and the geographical books of the sixteenth century, the answer is obviously not, and the reason why not tells us something very important about Renaissance learning—that it had not freed itself from the authority of the past.

Geography, like all other forms of learning, existed in books: If one wished to learn about foreign countries, one studied the texts by whatever ancient authorities one could find. In the sixteenth century people were looking at the world with new eyes, and Europeans were sailing in seas whose very existence had been unknown before. But this did not yet mean that experience suddenly became the only test of what was true. If ancient authorities said that monsters existed, then they probably still did. If they had not been found in Asia, perhaps they existed in Africa or in America, and there are several cases in which the site of legends was actually moved in this way during the age of European exploration. The process by which experience became the only test of truth was a long one, and that was one of the reasons why the scientific revolution occurred so long after the other intellectual changes which we call the Renaissance.

We can sum this up by saying that in the Renaissance, society was dynamic, but learning was static, and the persistence of monsters on maps is one small but important example of this.

Another wonderful sea monster from Munster's *Cosmographia*.

This picture of Noah's Ark from the *Nuremberg Chronicle* gives an idea of shipbuilding techniques in the Renaissance.

Science from the Earth:
Georgius Agricola

Georgius Agricola, pioneer of metallurgy and geology.

The Renaissance discoveries in the applied sciences were embodied in a number of textbooks that became landmarks in their fields and in the history of printing. One of the most important of these early books in the history of technology was the great treatise on mining and metallurgy, *De re metallica* by Georgius Agricola, published in 1556. Agricola (1494–1555) studied medicine at universities in Italy before returning to his native region. He began by studying the medical ailments of the people who labored in underground caves and chambers, but his interest broadened into geology itself, and the techniques for extracting and refining the metal ores. For 25 years he gathered the materials for a series of books, culminating in *De re metallica,* which was ready for the press when he died and appeared posthumously.

In this book and in his other works Agricola approached the mysteries of the Earth in a firmly empirical spirit. "Those things that we see with our eyes," he wrote, "and that we understand by means of our senses are more clearly to be demonstrated than if learned by means of reasoning alone."

He rejected many traditional beliefs about the nature of the Earth. He did not believe that the world had been formed complete at one moment in time, as the Bible teaches. Nor did he accept the medieval idea that metals grew from seeds, as plants did, in the bowels of the Earth. He denied the astrological doctrine that each of the metals was somehow related to one the planets—silver to the Moon, iron to Mars, gold to the Sun, and so on.

Astrologers explained these affinities by claiming either that the metals had been formed by the influence of the planets, or even that the metals had somehow been carried to the Earth from those planets. Some authorities had even speculated that minerals might be male or female. Agricola reasoned correctly that most of the substances in the Earth had been formed by the action of fire and water, and that minerals had somehow been precipitated from liquid solutions. He also guessed that rocks and minerals were <u>metamorphic</u>—that new substances emerged as old ones were subject to pressure or fire, or were dissolved in water.

OBJECTIVE TESTS

One of Agricola's principal aims was to reduce to order the many terms that were used for one substance. He listed five tests that would produce more unified descriptions of minerals: color, weight, transparency, solubility, and texture. At that time, only six metals were recognized—

gold, silver, iron, tin, lead, and copper, but Agricola argued that mercury should be added to the list, and he thought that other, undiscovered metals certainly existed.

Using these tests, he compiled a catalog of some 600 substances, including minerals, stones, metals, and "congealed juices," that is, salts and sulfurs. He described the places where these substances typically occur and illustrated the shapes followed by veins of ore. In addition to this theoretical part of the book, Agricola went on to give detailed descriptions of the techniques for extracting metals from the ground and purifying them.

PRECISE ILLUSTRATIONS

One of the reasons for the book's lasting fame was the many hundreds of woodcut pictures that Agricola commissioned to illustrate his text. He showed the different kinds of winches and power trains used to bring material to the surface, pumps to keep the mines dry, and the earliest pictures of railroads—trucks that were pushed or pulled along wooden tracks, their wheels kept in place by grooves or guides.

In the refining processes he showed innumerable kinds of furnaces, bellows, heat-proof vessels of iron or glass, molds, and scales to test the purity of the finished product. He said: "I have hired illustrators to delineate their forms, lest descriptions that are conveyed by words should either not be understood by men of our own times or should cause difficulty to posterity." We are indeed fortunate to possess such a rich picture of sixteenth-century technology.

Agricola's text is a landmark in the history of science for two reasons. First, it is rich in empirical detail, in information drawn from firsthand experience; second, he was prepared to use this experience as a basis for new theories about the composition of the Earth. In both these respects he represents a new approach to the richness of nature. Clearly Agricola was not able to construct a new science of geology single-handedly, but he does stand at the beginning of modern geology. His science was firmly rooted in experience—in the field and in the workshop—and not in the mere repetition of ideas drawn from other books.

Agricola's was not the only book of its kind from this period, although it was the fullest. In 1540 a text had been published by an Italian expert, Vanoccio Birunguccio (1480–1539), called *De la pirotechnia* ("On Pyrotechnics"). It dealt with the smelting and casting of metals, especially the making of guns and other weapons, and the ways of preparing gunpowder. Like Agricola, Birunguccio dealt with processes and equipment that were also the concern of the alchemist. But the great difference is that these authors stripped away the dreams and the occult jargon of the alchemist, and were content to describe physical processes and to try to give systematic accounts of what happens when substances are heated, melted, purified, or combined.

An English translation of Agricola's great Latin text *De re metallica* was published in 1912 by Herbert Hoover, the future president of the United States (1929–33), himself a mining engineer.

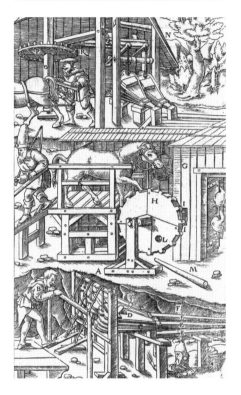

Above: Air being pumped into mines, and minerals being raised by treadmills, from Agricola's book.

Below: Gunpowder being prepared from saltpeter, sulfur, and charcoal; the devil guides the scientist who is creating this murderous mixture.

Renaissance Biology:
The First Naturalists

The sixteenth century saw major advances in descriptive biology by writers whose analysis of the world of plants and animals went far beyond the medieval herbals and bestiaries (see Volume 2, page 64). There were several reasons for a new approach to botany and zoology at this time. First, the discovery of the New World greatly enlarged the range of flora and fauna known to Europeans, and made it desirable to try to describe them all systematically. Second, the painters of the Renaissance were producing splendid images of plants and animals that made the illustrations in scientific manuscripts appear crude and inaccurate. Third, the advent of printing created the opportunity to set down firsthand observations of nature and circulate them among other scholars for the purposes of study and criticism.

The important early botanical works were all German, then Italian scholars followed. First came *Herbarum vivae eicones* ("Illustrations of Living Plants") by Otto Brunfels, published in three parts between 1530 and 1536. Brunfels (1489–1534) did not attempt any systematic classification of plant types, but the importance of his work rests on the pictures of 238 plant species. They were drawn from life, where earlier plant book had simply copied one another—some of the pictures show plants that are wilting or damaged, leaving no doubt that they were faithful records of real specimens. However, Brunfels's great limitation was that he described only plants that he knew already from earlier texts of classical botany.

Botanical artists at work from Fuchs's *De Historia Stirpium* ("History of Plants"), 1542; Renaissance naturalists rejected mere copying and drew their specimens from nature.

DETAILED DESCRIPTION

A new approach appeared in the second of these German pioneers, Jerome Bock, whose *Neu Kreutterbuch* ("New Plant Book") was published in 1539. The first edition contained no pictures, although it was later reprinted with

Heinricus Füllmaurer.   Albertus Meyer.

illustrations. Bock (1501–66) presented around 700 plants, with descriptions that were precise enough to identify them in the field. He attempted to group them into types, using the shape of the plant, the petals, or the leaves as criteria. Bock, like all his contemporaries, did not know enough about plant physiology to arrive at true classifications based on reproductive characteristics and seed types, but his was an early attempt to bring order into the multitude of plant types.

ROSA

Less systematic than Bock was Leonhard Fuchs (1501–66), whose book *Historia Stirpium* ("History of Plants") in 1542 was beautifully illustrated and gave careful descriptions of the habitat and characteristics of hundreds of plants. The fuchsia was named after him, although he did not describe it since it was imported from America after his death.

Together, Brunfels, Bock, and Fuchs may be said to have revolutionized botany by emphasizing firsthand field study and avoiding mere copying from ancient authorities.

The new art of botanical description and illustration reached its highpoint in 1544 with the publication of Pier Andrea Mattioli's revision of Dioscorides' *De materia medica* ("The Substances of Medicine"), with hundreds of faithful plant pictures drawn from life.

Above and Below: Leonhard Fuchs's *Historia Stirpium* included beautiful, accurate illustrations of many plants, including the rose (above) and cherry tree (below).

CAPTURING ANIMALS

The animal kingdom, too, had its Renaissance pioneers and illustrators, but a wild animal is obviously less passive as a subject than a plant, and certain misconceptions or legends about animals tended to persist for longer. The best-known example of this is the belief that the rhinoceros had segmented armor plating, something that began with Dürer's woodcut print of 1515 and was recopied for nearly 200 years.

One of the most original of early zoologists was the French scholar Pierre Belon (1517–64), who traveled widely in Europe and the Middle East, studying mainly birds and fish, finding many of the latter in exotic food markets. Belon's great interest lay in comparative anatomy and speculating on relationships between species. He dissected cetaceans

CERASVS

(whales and dolphins), and the milk glands that he found convinced him that they were sea-dwelling mammals. He was struck by the underlying similarity of forms in different species, and in his book *Histoire de la Nature des Oysseaux* ("Natural History of Birds"), published in 1555, he drew a human skeleton beside that of a bird, showing that these widely different species shared some similarities of general structure. Belon was the first to divide the bird kingdom into major groupings such as raptors, waders, web-footed birds, nocturnal hunters, and so on.

CLASSIFYING ANIMALS

In the same year as Belon's study of birds there was published the most monumental of the new zoology books, the *Historia Animalium* ("History of Animals") by Conrad Gesner. Gesner (1516–65) was a native of Zurich in Switzerland and was a physician and classical scholar as well as a naturalist. He was perhaps the first person we know of to climb mountains for fun, describing his ascent of Mount Pilatus near Lucerne. His great book was a veritable encyclopedia in five volumes, showing thousands of creatures, and it remained a source-book in zoology for two centuries to come. Gesner used general categories such as quadrupeds, oviparous animals, and so on. Strangely, alongside the hundreds of carefully drawn pictures of real animals Gesner described mythical creatures—dragons and mermen—as though he were not quite

CONRAD GESNER (1516–65)
- Naturalist and physician.
- Born in Zurich, Switzerland.
- Early education in the house of his great-uncle, a collector and grower of medicinal plants.
- Schooled in Strasburg, Bourges, and Paris.
- 1537 Became professor of Greek at Lausanne Academy aged 21 and published his first work, a Greek-Latin dictionary.
- Settled in Zurich, 1541. Became professor of philosophy and natural history at Zurich University.
- Spent the rest of his life in Zurich working as a doctor and teaching Aristotelian physics at the Collegium Carolinum.
- 1554 Became city physician of Zurich.
- 1545–49 Published a compilation of all the known books published in Greek, Hebrew, and Latin with a summary and criticism of each.
- 1548 Published an encyclopedia which attempted to convey all the knowledge of the world, initially in 19 volumes.
- 1549 Volume 20 on theological thought published; Volume 21 on medicine was never finished.
- 1551–58 Published *Historia Animalium* ("History of Animals")—a descriptive work of all the known animals in the world.

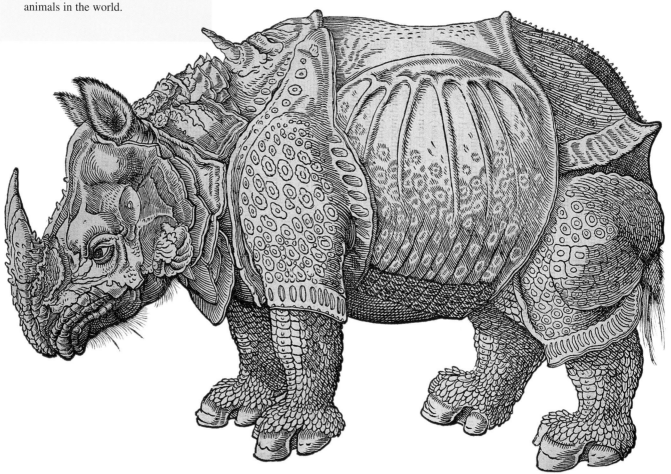

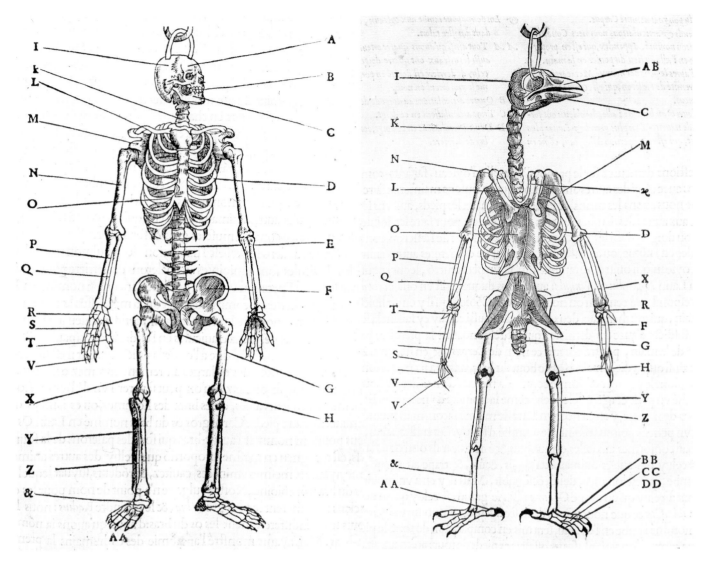

sure if these things existed, but included them anyway because he did not want his work to be incomplete.

These early naturalists were all looking at nature with new eyes, although they could not entirely free themselves from the past. They were looking for groups, patterns, and relationships among living forms, seeking the fundamental structures beneath the diversity of nature. Many legends still persisted, and some were used to explain what could not be discovered by observation. Even the division between the animals and plant kingdoms was not absolute. Jerome Bock, for example, was puzzled by the way orchids propagate and suggested that they were the offspring of birds. John Gerard's (1545–1612) famous herbal *The Historie of Plants,* published in 1597, applied the same idea to the brant goose—suggesting that it was hatched from barnacles. The same credulity was still evident about the human world, for travelers' tales from the New World showed that belief in giants, monsters, and half-human species was still very much alive. The printed works, such as those by Fuchs or Gesner, were treasure houses that delighted the eye and stimulated the mind, but the systematic analysis of form and function in biology lay in the future.

Above: Skeletons of man and bird by Belon, 1555; Belon realized that the structures showed general similarities, but he did not suspect an organic relationship.

Opposite: Gesner's rhinoceros: When virtually no European had ever seen an African animal, myths would grow up about it—such as the rhino's armor plating.

Medicine and Mysticism:
Paracelsus

It would difficult to imagine a greater contrast to the observational science of Agricola than that of his exact contemporary, Paracelsus, although he, too, was a physician who grew up in a mining region and produced theories about the composition of the Earth.

Paracelsus (1493–1551) was a native of Switzerland who combined original research in scientific medicine with elaborate and obscure theories concerning the transcendent forces that he believed were at work in nature.

After studying medicine at universities in Germany and Italy, Paracelsus traveled far and wide as a military surgeon, during which time he gathered material for his book *Grosse Wundarzney* ("Great Treatise on Wounds") of 1536, which made him famous. He was an intellectual rebel, rejecting traditional medical ideas, and on one occasion he publicly burned the works of Avicenna, the great medieval Islamic physician. He pioneered many original medical treatments: He demonstrated that the disease that afflicted miners, silicosis, was caused by inhaling particles of coal dust; he diagnosed that the problem of goiter was caused by drinking water low in iodine; and he showed that syphilis could be treated with small doses of mercury.

NEW THINKING

Central to Paracelsus's quarrel with orthodox medicine was his rejection of the traditional doctrine of the humors, which taught that all disease was caused by the imbalance of the four elements in the body (see Volume 2, page 4). In this scheme diseases were not classifiable as functional disorders, for each manifestation of disease was separate and personal to the patient. Against this Paracelsus argued that diseases arose from direct causes in the external world that they were agents that invaded the body and threatened its life or well-being.

This sounds remarkably modern in its approach, but what were these agents in Paracelsus's view? Strangely, the causes of natural disorders turn out to be supernatural in character: Paracelsus explained that they were poisons of various kinds emanating from the earth, the air, and even the stars; he said that they were "astral bodies" that used the elements of this world as vehicles for their actions. But just as these poisons emanated from the realm of nature, so too did their cures. For example, in the case of syphilis he believed that mercury worked by expelling poison salts from the body, but it

Paracelsus, physician, scientist, and mystic.

PARACELSUS (1493–1541)
- Originally named Philippus Bombastus von Hohenheim.
- Alchemist and physician.
- Born in Einsieden, Switzerland.
- Possibly graduated from Vienna University and took his doctorate in Ferrara.
- 1526 Started lecturing on chemistry at Basel in German rather than Latin and allowed barber surgeons to listen, displeasing the authorities, who wanted education to remain elitist.
- In his lectures he emphasized observation and experiment; taught that disease was caught by the body, not generated from within.
- By 1528 had to flee Basel.
- The first physician to recognize congenital syphilis; made particular studies of tuberculosis and silicosis.
- Introduced sulfur, mercury, lead, and laudanum into Western medicine.
- 1541 Finally settled in Salzburg just before he died.

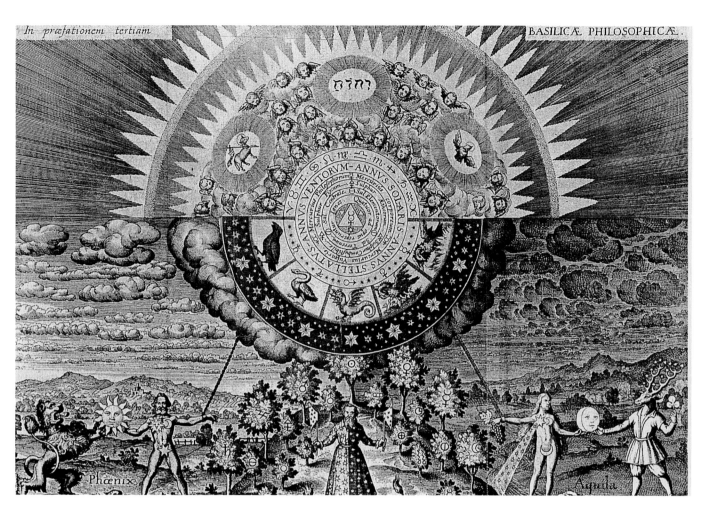

was the celestial virtue of the planet Mercury that was working through the medicine. This healing effect was possible because all nature was linked in a system of correspondences: Minerals, plants, animals, and stars all shed their influences on humans, sometimes for good and sometimes for ill.

The Paracelsian cosmos: Animals, plants and minerals are linked in a system of relationships—the man and woman have the natures of the Sun and Moon, and are chained to the heavenly sphere.

HIDDEN FORCES

Paracelsus imagined an invisible world teeming with spiritual forces that, however, operated in natural, analyzable ways. Thus the whole realm of nature was considered to be alive and its parts interrelated. This belief involved both alchemy and astrology, but of a rather special kind in which physical objects, such as minerals and stars, possessed essences or spirits. It was the aim of the scientist to understand and control these spirits. This was a semimystical vision of nature that, to modern eyes, appears closer to magic than to science, and it was typical of the Renaissance "magus," or wise man. There were many self-proclaimed "magi" in the Renaissance, perhaps the best known being the fictitious character Dr. Faustus, who exhausted all conventional wisdom and then sold his soul to the Devil in exchange for knowledge of nature's inner secrets. Paracelsus's thought represents a strange blend of physical science and occult beliefs that shows clearly the way that Renaissance scientists were eagerly exploring new theories of nature.

Science and Magic:
The Philosophy of Hermeticism

Cornelius Agrippa, the notorious German occultist who was the model for Dr. Faustus.

There was one aspect of Renaissance science that appears to us to be very strange and very different from the mathematical and empirical sciences such as navigation, military engineering, or biology. This was the magical or occult tradition. In part, it was descended from medieval beliefs, for example, in the fields of astrology and alchemy (see Volume 2, pages 48 and 55). But this kind of science received a strong, renewed impetus in the Renaissance through the discovery of a number of Greek documents that were translated into Latin in the 1470s by the Florentine scholar Marsilio Ficino. These documents were philosophical and scientific texts that claimed to contain the teachings of the ancient Egyptian deity Thoth, who was the god of wisdom and language. The ancient Greeks had known these texts, and they had identified Thoth with the Greek god Hermes; this philosophy, therefore, became known as Hermeticism.

MAGICAL PHILOSOPHY

Knowledge of these texts and their ideas had been lost to the West during the Middle Ages; but when Ficino's translations began to circulate, they aroused enormous interest. Thoth was believed to have been a real person—half human and half divine—who had lived in Egypt almost 5,000 years before, and these texts were seen as offering an alternative account to that given in the Bible of creation and of humankind's nature.

Central to the Hermetic philosophy was the idea that the universe was a unity. At its creation it had been endowed with energies that are also spoken of as personal spirits or demons. The universe had been created by an all-powerful God, who now used these energies or spirits to govern the world of nature.

Humankind was given a high place in this scheme. As in the Bible, humans had been made in the image of God, but this was taken to mean that they could become like God. Humans could learn to control the forces of nature through alchemy, through astrology, and through the study of magical formulas using plants and minerals.

All nature was thought to be animated by spirits or demons that humans could learn to command. This was an exciting, intoxicating vision, and many scholars in the Renaissance devoted themselves to this magical philosophy of nature. One of the most famous was Cornelius Agrippa (1486–1535), who published an encyclopedia of

SVSCIPITE·LE
PITE·LE·GES
O·LI EGIP
CTE
RAS

DEUS OMNIUM CREATOR
SECU M DEUM FECIT
VISIBILEM. ET. HUNC
FECIT PRIMU·M·ET SOLLUM
QUO OBLECTATUS EST ET
VALDE·AMAVIT PROPRIUM
FILIU·M·QUI·APPELLATUR
SANCTUM. VERBUM.

HERMIS· MERCURIUS·TRIMEGISTUS
CONTEMPORANEUS · MOYSI

magic called *De occulta philosophia* ("Philosophy of the Occult"). It gave instructions for drawing down the powers of nature by astrological talismans, by the magical properties of numbers, by reciting spells, or by burning or distilling mixtures of minerals or herbs. Agrippa was a possible real life model for Dr. Faustus, the magician in Christopher Marlowe's play who gives his soul to the devil in return for unlimited knowledge and power over nature.

All this has a very pagan sound, and it was condemned by many church authorities. But the reason for their condemnation was always that these practices were forbidden to humankind. The truth of the Hermetic ideas was not questioned, for they represented a consistent philosophy of nature, and God was acknowledged as the ultimate source of all nature's powers.

Hermes-Thoth, the deity who supposedly taught science and philosophy to the ancient Egyptians. Texts attributed to Hermes greatly influence Renaissance scientists such as Bruno and Kepler. This mosaic picture is in Siena Cathedral, for Hermes was believed to have been a prophet of Christ.

Renaissance scholars were particularly impressed by one special feature of the Hermetic writings, namely, that they appeared to prophesy the coming of Christ. They spoke of a "Son of God" who would renew the ancient teachings. For this reason Hermes became a figure whose authority was recognized even by Christian theologians, and images of Hermes were placed in churches. Later, in the seventeenth century historians discovered that the Hermetic writings originated in the second century A.D. and not from ancient Egypt, so that the "Son of God" passages were not prophecies at all.

QUESTIONING RELIGIOUS BELIEF

Probably the most famous Hermetic philosopher was Giordano Bruno (1548–1600) who was burned at the stake in Rome for heresy. Bruno traveled through many European countries, lecturing and writing on the Hermetic philosophy. He was also a follower of Copernicus: He had stated his belief that the Earth was a planet, and that the universe was infinite. This could be seen as evidence of atheism, for an infinite universe could not be created by anything outside itself. Equally important, the idea that the Earth is just one planet in an infinite universe called into question the religious belief that humankind was the crown and the purpose of God's plan. Bruno was asked to retract a number of such heretical statements found in his works, but he refused. Bruno is often seen as a martyr for the new science, and perhaps he was; but the detailed documents of his trial were destroyed long ago, and we do not know exactly on what grounds he was condemned. He is also said to have denied the divinity of Christ, so precisely where his heresy lay, whether in his scientific or in his religious ideas, is now impossible to say.

Tomasso Campanella, priest, poet, revolutionary, and advocate of astrology and magic.

RESISTANCE MOVEMENT

Another great Hermetic philosopher was the priest, philosopher, and poet Tomasso Campanella (1569–1639), who, like Bruno, suffered terribly for his beliefs. Campanella became involved in a resistance movement by the people of the Naples region of Italy against the Spaniards who then ruled over them. He was captured and tortured, and kept incarcerated for a total of 27 years, but from his prison cell he poured out a stream of philosophical works.

His most famous book was called *The City of the Sun,* a vision of an ideal city governed by a priesthood devoted to Hermetic magic. These leaders would draw down the astrological powers of the stars to ensure that the people of the city lived in harmony. At the center of the city was a temple of the Sun surrounded by seven circular districts representing the seven planets, for Campanella, too, was a Copernican. The walls of the city would be decorated with portraits of great prophets and philosophers of the past, including Moses,

Hermes, and Christ. The ideals of the "City of the Sun" were those that had inspired Campanella's political activity, and he never abandoned hope that a great leader would begin a reform of civil and religious life based on Hermetic principles.

After his eventual release from prison Campanella demonstrated his magic to none other than the pope himself. Pope Urban VIII was a scholar who had a deep interest in astrology. He had become convinced that a forthcoming eclipse would cause his death. Campanella arranged a private ceremony for the pope in which lamps symbolizing the planets were arranged to form a substitute heaven, and with magical incantations, the substitute eclipse passed off without harming the pope.

A Renaissance occultist raising a demon, standing safely within a magic circle in the manner described by Agrippa.

THE PROMISE OF POWER

In the sixteenth century, and well into the seventeenth, no distinction was made between science and what we now call magic. Both were part of a philosophy of nature because both sprang from a belief that nature was governed by laws that were discoverable by humans. The difference between Hermeticism and other approaches to science was the emphasis on humans' ability to command nature's forces. The great weakness of Hermetic magic was that its adherents wanted to control nature before they really understood it.

The magical formulas, the alchemical and astrological doctrines were seized on because they promised power to the person who mastered them. They were never subjected to empirical tests; and if they failed to deliver what they promised, it was because the practitioner had not performed them correctly, not because they were false. Hermeticism retained a strong influence over many great scientists of the Renaissance such as Kepler and, perhaps, Copernicus, and even in the age of Newton. In a sense, the Hermetic vision of humans as imitating God by learning the secrets of nature's powers is still alive today, for it can be seen as a prophecy of the powerful modern technologies, by which humans are controlling and reshaping their environment and their lives.

Science as a Renaissance Profession: Girolamo Cardano

One of the most intriguing figures in sixteenth-century science was the Italian Girolamo Cardano (1501–76), who achieved fame as a mathematician, physician, and astrologer. He is also a figure who comes to life to an exceptional degree because he was involved in a number of very public quarrels, and he wrote a very revealing account of his own life. He tells how he survived enormous handicaps and adversities, and how miraculous powers were given to him to solve difficult intellectual problems. He wrote encyclopedic works that are a strange mixture of superstition and original scientific ideas.

YOUNG PRODIGY

The illegitimate child of a scholarly father, Cardano was weak and sickly, ill-treated by his parents, and near death on many occasions. On entering the University of Pavia, he knew no Latin, but he bought a Latin book for its sumptuous binding and suddenly found he could read the language perfectly. He studied medicine and later practiced as a physician in a small town near Padua. He had a natural gift for mathematical calculation that was fed by his addiction to gambling. Chess and dice games fascinated him, and he was always calculating the possible outcomes.

The astrologer with a client, filling in the celestial positions on a horoscope chart; unrealistically, the Sun, Moon, and stars are all shining together.

At this period Cardano was both poor and ambitious, and he decided that astrology might provide the best hope of fame and wealth. He had studied astrology as part of medical theory, and in 1534 he published a "Prognostication," a series of predictions for the coming year and for the longer-term future that enjoyed considerable success.

Cardano was critical of contemporary astrologers for relying on precalculated tables of celestial positions that were often inaccurate and never studying the heavens in person. He also began to publish works on mathematics, and in 1539 he met Niccolo Tartaglia, another leading mathematician, who informed him that he had discovered formulas for solving cubic equations in algebra, which until this time had been considered impossible. Cardano persuaded Tartaglia to reveal the method to him, at the same time promising never to publish it. But Cardano broke his promise and in 1545 published a book called *Ars Magna* ("The Great Art"). It broke new ground in algebraic practice, including the method of solving cubic equations. This provoked the first of Cardano's public controversies, a fierce quarrel with Tartaglia about the priority in this discovery.

A Renaissance astrologer at work with his books and instruments.

FLAWED GENIUS

Cardano revealed an enormous amount about his life and personality, not attempting to conceal his defects. His torso was thin, and his health was always bad. He suffered diseases of the lungs and skin, fevers, and insomnia. His voice was shrill, and his coordination was bad. He was quarrelsome, fell out with his friends, and became involved in many lawsuits. His faculties were strangely acute, and he believed in all kinds of omens. When someone was talking about him, however far away, he could hear a whispering in his ear—if they were praising him it would be in his right ear, if damning him, in his left. When his son was about to make a disastrous marriage, he was forewarned by a small earthquake that shook his house. His own marriage had been preceded by a fire that lit itself in his hearth—something he took to be a good sign. Just before he was forced to abandon public teaching, his dog tore up his lecture notes. He was fond of small animals, and his house was infested with lambs, rabbits, and birds that he believed brought him good luck. He was wildly eccentric, but he believed that special powers of intellect had been granted him to compensate for his strangeness, and that some guardian spirit watched over him and gave him signs foretelling the future.

The mathematician Niccolo Tartaglia, with whom Cardano quarreled bitterly.

A STEP TOO FAR

In 1543 Cardano took up the professorship of medicine at the University of Pavia and soon became as famous in the world of medicine as he was in mathematics. He was consulted by wealthy nobles and traveled to many parts of Europe, as far afield as France, Scotland, and Denmark, to treat them. It may have been his experience of court circles that gave him the

GIROLAMO CARDANO (1501–76)

- Mathematician, philosopher, naturalist.
- Also known as Hieronymus Cardanus and Jerome Cardan.
- Born in Pavia, Italy.
- 1543 Became professor of medicine at Pavia.
- Invited to Scotland to treat the archbishop of St. Andrews in 1551. Then journeyed to London to cast a horoscope for Edward VI.
- Prolific writer, he wrote over 200 papers about everything from astrology and philosophy to music, math, and physics.
- 1562 Became professor of medicine at Bologna.
- Cardano's formula solves cubic and quadratic equations. He explained how in a work on algebra entitled *Ars Magna* ("The Great Art").
- Imprisoned by the Inquisition for heresy in 1570, recanted the following year and was released.
- 1571 Went to Rome and was granted a pension by Pope Pius V.
- 1576 Finished his candid autobiography and died soon after.

idea for a new book on astrology that was to have fateful consequences on his life. His idea was to publish "celebrity" horoscopes of famous contemporaries and people from the past, including King Henry VIII of England, the Emperor Charles V, Cicero, Nero, and Michelangelo. These were not predictive horoscopes since Cardano had already studied the characters and careers of these people, but he offered them as illustrations of the principles of astrology—how the influence of stars explained the destiny of these exceptional figures.

Published in 1552, these horoscopes increased Cardano's fame, but plunged him into crisis, for he was audacious enough to include the horoscope of Christ himself. Cardano gave the positions of the heavenly bodies for midnight on December 24 in the year 1 B.C. and interpreted them by saying that they clearly predicted the birth of a noble, wise, and loving prophet who would suffer an unjust death at an early age; in other words, Christ's life, like that of any other human being, was subject to the influence of stars. This was a very dangerous statement to make, for it contradicted the theological doctrine that Christ's incarnation and self-sacrifice were divine miracles.

The notorious horoscope of Christ published by Cardano in 1552.

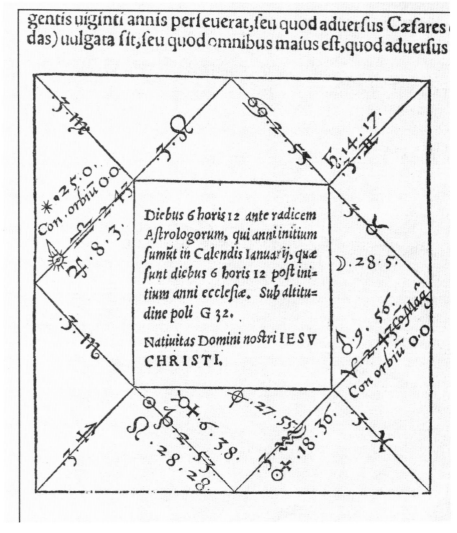

Cardano became the victim of his own recklessness. He was arrested by the Inquisition for blasphemy, thrown into prison, and ordered to retract his statements. Cardano submitted and was released, but he was deprived of his teaching post and forbidden to publish ever again. He was ordered to live in Rome for the rest of his life under the watchful eye of the church authorities.

Meanwhile, a fresh tragedy overwhelmed him: His eldest son was convicted of murdering his wife (she whose marriage had caused the earthquake in Cardano's house) and was executed in spite of all Cardano's efforts to save him. This, Cardano was convinced, was due to the many enemies he had made through his writings.

In his later years he composed a number of works that appeared only after his death. They included one of the first books on the mathematics of probability, about which Cardano had learned a great deal from his gambling days, and his highly revealing autobiography.

Girolamo Cardano, physician, mathematician, and astrologer, whose scientific career was ended by his own recklessness.

Cardano was undoubtedly a gifted mathematician, and he was a tenacious personality, determined to use his intellect to overcome his huge handicaps. But he was also a showman who saw that the display of special knowledge could bring him wealth and fame. He made a career out of science, using the medium of printing to reach other scholars, but still more to impress educated laymen. In the end he overreached himself, and for the last years of his life he was silenced. The story of his life gives a human dimension to the world of sixteenth-century science, showing how that science was in transition from superstition to rational knowledge.

Renaissance Medicine:
Medicine before Vesalius

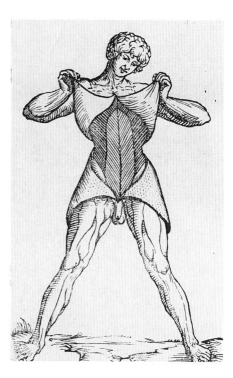

Above: An artistic but vague scheme of the dissected torso muscles, by Berengario da Carpi, 1521.

Below: Pres-Vesalian anatomy was often imprecise and illustrated what was believed to exist, not what the experimenter actually saw.

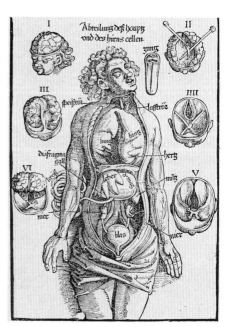

The evolution of medicine during the Renaissance was partly related to the artistic rediscovery of the human body. Painters, sculptors, and philosophers commended the perfection of the human form after centuries during which medieval thinkers had, for religious reasons, denigrated the body in favor of the soul. A new understanding of the importance of anatomy is evident from around the year 1500. Formerly anatomy had been regarded as secondary to the functioning of the body, which was always explained in terms of the four humors (see Volume 2, page 4).

The theory of the four humors was challenged by the appearance in Europe of the disease syphilis, with its horrifying symptoms. It is generally thought that syphilis was brought back from America by returning explorers in the 1490s, since that is when a virtual epidemic of the disease spread from the Mediterranean countries. Historians of medicine have recently disputed this, arguing that it had been present in Europe before, but had been less virulent or had been confused with leprosy. Whatever the truth may be, contemporaries viewed it as a new curse, and they were forced to ask where it had come from. Its obviously contagious nature could not be reconciled with the classic notion that all disease results from the imbalance of the humors in the individual. The Italian physician Girolamo Fracastoro (around 1478–1553) suggested that the origin of the disease lay in "seeds" that could be carried from person to person, and that grew inside their victims. This was the first statement of a germ theory of disease.

MEDICAL TEXTBOOKS

Printing had an enormous impact on medicine, most obviously in the dissemination of anatomy texts with their accompanying illustrations. The first printed anatomy book was the *Fasciculus Medicinae* ("Bundle of Medical Papers") of 1491 by a physician of Vienna named Johann von Kircheim (1455–90), commonly known as Ketham. The drawings in this text are extremely crude, and they illustrate preconceived schemes of the bodily functions based generally on the theories of Galen rather than firsthand observation. Ketham's text was later illustrated with some general pictures of the doctor at work: The picture of the dissection room, for example, makes it clear that the dissection itself was carried out by an assistant, while the doctor lectured from above on what the student should see. The veneration for the classical medicine of Galen went so far that Jacobus Sylvius, a professor in Paris, declared in 1555 that if what the eye saw during a dissection did not conform with Galen's description, then the fault lay not with Galen but with the corpse; modern human form, Sylvius suggested, might have changed since the classical era.

One of the earliest works to report personal observations drawn from dissection was the *Commentaria* ("Commentary") by Berengario da Carpi, 1521. This work contains the first references to features such as the vermiform appendix and the thymus gland, and da Carpi (1460–1530) denied the presence in the human brain of the *rete mirabile,* a network of veins that Galen mistakenly described because he found it in dissected animals. Some of da Carpi's pictures bear a strong resemblance to the anatomical drawings of Leonardo da Vinci, suggesting that he may have seen Leonardo's work.

Model of the anatomical theater of Fabricius in Padua.

One of the most puzzling areas of the body was the brain, and many theorists tried to locate the different senses and mental faculties, such as memory or reason, in separate sections of the brain.

PRACTICAL MEDICINE

Certain fields of medicine tended to be regarded as crafts, as areas where practice was more important than theory. This was true of obstetrics and surgery, which were not necessarily part of the academic physician's role. If a good, well-illustrated text was printed in this field, it tended to retain its authority for many years, or even decades. Eucharius Rosslin published the first book on childbirth, called *Buchlein der Schwangeren Frauen* ("Handbook of Pregnancy"), in 1513. It became so popular that it was translated into French, Italian, Spanish, Dutch, and English (entitled *The Byrth of Mankynde,* 1540). Likewise surgery, in its most common form of battlefield treatment, was covered in Hans von Gersdorf's *Feldtbuch der Wundarzney* ("Field book of Wound Types"), 1517, which described how to deal with the dreadful wounds caused by saber cuts or cannonballs. Gersdorf devised many instruments himself and many original and effective treatments, although the pictures in his book are bloodcurdling in the extreme.

SURGERY IN THE FIELD

The most famous surgeon of the sixteenth century was the Frenchman Ambroise Paré (1510–90), who had no formal education, but who had learned his craft on the battlefield. He discovered that the drastic treatments then used for wounds, such as cauterizing with hot iron or oil, did more harm than good and substituted salves of egg whites or creams, or other natural antibiotics. Paré published an authoritative text called *Five Books on Surgery,* 1572; but when he ventured his opinions on strictly medical topics, he was condemned by the Paris medical faculty as a mere surgeon who had no right to encroach on medicine proper. Paré also designed a number of artificial limbs to replace those lost on the battlefield.

THEORY OF CIRCULATION

One of the most significant new ideas in physiology, challenging Galenic theory, came from a man who was an unorthodox physician and theologian, the Spaniard Michael Servetus (around 1511–53). Servetus discovered the "lesser circulation" of the blood, also known as the pulmonary transit. This refuted the traditional idea that blood permeated between the two sides of the heart, but argued instead that it traveled from the right side to the lungs, where it acquired air before returning to the left side. This was halfway to William Harvey's later discovery (see Volume 5, page 44). Servetus announced his discovery in the very strange context of a theological work in which he likened the entry of air into the blood to the action of God in breathing the soul into Adam's body, as described in the book of Genesis.

This book had a tragic outcome, for Servetus had become sickened by the religious strife that was plaguing Europe, and in this same book he advanced <u>antitrinitarian</u> views. News of these ideas reached Calvin and the church authorities in Geneva; and when Servetus very unwisely visited the city, he was arrested, tried for heresy, and burned at the stake. Servetus's ideas had no influence on later physiology, mainly because almost all the copies of his book were also burned.

MEDICINE FROM PLANTS

A minor revolution of the sixteenth century that affected medicine was in botany and pharmacology. New plants in their thousands were brought back to Europe from the Americas and the Indies, while botanists and artists strove to give more precise descriptions

The human brain, analyzed to show where senses such as vision, taste, and smell were supposed to be located, by Johannes Dryander, 1557.

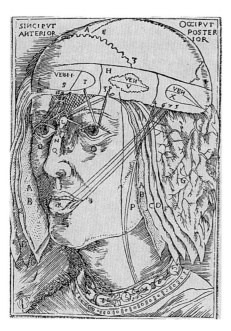

Fifteenth-century medical students at work on a dissection.

of those species that were already known. This was most important for the preparation of drugs, since gathering the wrong plants could lead to disasters in treatment. The first professorships in botany were established at universities in Italy in the 1530s, while scholars for the first time began to plant their own herb gardens to cultivate and study the new plants.

Among the substances brought from America were cocoa and tobacco, both of which were claimed as wonder drugs, while from the east came opium, about which one sixteenth-century doctor exclaimed, "Among the remedies that it has pleased almighty God to give to man to relieve his sufferings, none is so universal and so efficacious as opium." Most of its effects may, no doubt, be attributed to its pain-deadening quality. Apothecaries became established in their own professional <u>guilds</u> in the late sixteenth century.

Renaissance Medicine:
The Reforms of Vesalius

Renaissance medicine achieved its first highpoint with the career of Andreas Vesalius (1515–64). Vesalius, a native of Brussels, was an anatomist who did not develop any new theories about the workings of the body or the causes of disease. But he recognized that no true approach to these problems was possible until the structure of the body was correctly understood. Physicians must not simply find the organs, muscles, or nerves that Galen had said were present, but must study the body minutely with their own eyes. He showed his zeal for anatomy by robbing wayside gibbets for the bodies of executed criminals.

In 1537 Vesalius moved to Padua, Italy to teach at the university. In the following year he published his first medical text, but at this stage he could not free himself from the influence of Galen, and the pictures that he drew still showed some features inaccurately.

Portrait of Andreas Vesalius, who brought new standards of accuracy to anatomical study.

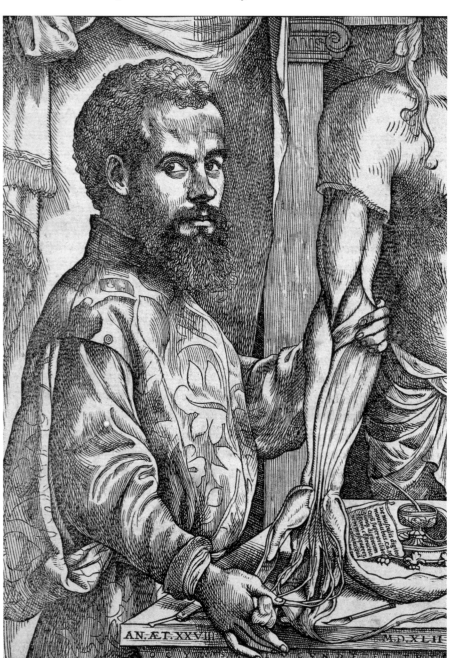

SEEING IS BELIEVING

He soon realized that he needed to be more rigorous in reporting what he saw. In a few years he became convinced that Galen had never dissected a human body, but had based his anatomical ideas on analogies drawn from animals. Galen taught that the body's central vein, the *vena cava*, had its source in the liver, but Vesalius saw it entering the heart. Galen asserted that the human jaw is formed of two bones. Vesalius showed this was true of many animals, but not of humans. In discussing the brain, Galen had claimed that sensation and motion were guided by an "animal spirit" that was gathered in the *rete mirabile* at the base of the brain, yet Vesalius plainly saw that no such organ existed in humans. He could find no trace of the minute valves that Galen said allowed blood to pass between the right and left sides of the heart. He also saw that nerves were not hollow tubes along which flowed some "vital fluid," as Galen had suggested.

Vesalius felt that he must embody his new, empirical approach in an atlas of anatomy that would set new standards of clarity and realism. He spent several years, therefore, working with a Flemish artist, Jan Stephan van Calcar (1499–around 1546), to prepare a series of anatomical illustrations that would revolutionize medicine. His great book, *De humani corporis fabrica* ("The Structure of the Human Body") was published in Basel, Switzerland, in 1543, and the superb illustrations of the human form stand as a landmark of printed art as well as of science.

The pictures were prefaced by the Vesalian principles that the eye was to be trusted above the authority of ancient writers, that the physician must personally learn to dissect, and that anatomy was the key to all progress in medicine. The text was not without errors, the weakest area being in female anatomy and embryology. Vesalius acknowledged this, explaining that he had only ever dissected three female bodies. He also declined to speculate on theoretical matters—for example, whether the brain or the heart was the seat of the soul—because there was no anatomical basis on which to make deductions.

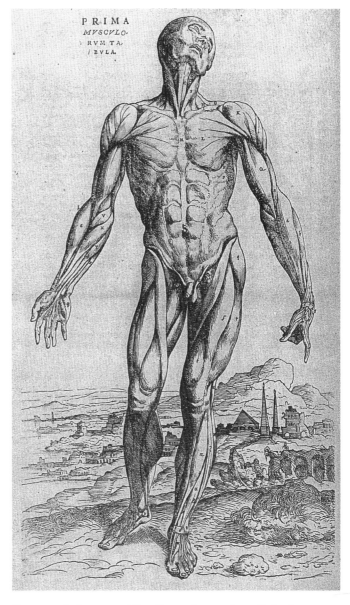

The frontal muscles of the human body, from Vesalius's *De humani corporis fabrica* ("The Structure of the Human Body") 1543.

THE BEGINNING OF OBSERVATIONAL SCIENCE

Although Vesalius cannot be credited with any single vital discovery, his work was revolutionary. For the first time, an ancient scientific authority, venerated for 1,500 years, had been tested against a body of empirical facts gathered by firsthand study, and that authority had been found to be badly wrong. Vesalius's pictures stand at the beginning of a new form of observational science in which later study and deduction are made possible by the availability of the scientific image. Traditionalists in the medical world resisted Vesalius's overthrow of the ancient authorities, but he inspired a new school of empirical anatomists who learned to trust their own eyes, and who discovered many new features in the body.

Renaissance Astronomy:
Before Copernicus

In the fifteenth century astronomy involved very little empirical or observational work. Its main function was twofold: to study and clarify the structure of the cosmic spheres, as described in the classical theory of Ptolemy, and to underpin the practice of astrology. For both of these purposes the astronomer's work consisted of mathematical calculation of celestial positions. Optical sighting devices did exist, but most astronomers and astrologers relied on precalculated tables of celestial positions and mathematical formulas. The most widely used of these tables were the so-called "Alfonsine Tables," which had been compiled in the middle of the thirteenth century (see Volume 2, page 46). Thus the astronomer of the thirteenth century was distanced from the direct study of the heavens by the rigid theory of cosmic structure that had been inherited from the remote past and by the use of precomputed data.

This situation began to change in the years 1460–90 with the innovative work of two German astronomers, Georg Peurbach (1423–61) and Johannes Muller (1436–67), who is always known by his Latin name of Regiomontanus. These two men decided to undertake a thorough revision of Ptolemy's great text the *Almagest*, first by taking a small number of very careful optical sightings of celestial positions, and then by working through his calculations. The *Almagest* was the great authority in medieval astronomy, but it is a long and technically difficult work, and most people knew it only in shortened versions, while no Western astronomers had ever rechecked its accuracy in this way. Peurbach and Regiomontanus then intended to compare their results with the data given in the Alfonsine Tables.

Peurbach died when the task was less than half finished, but Regiomontanus carried it through to completion by the year 1463.

Above: A model of the celestial sphere, from Regiomontanus's new edition of the *Almagest*, 1496.

Opposite: Regiomontanus, who revised and updated the data in Ptolemy's *Almagest*, and opened a new era in astronomy.

IOHANNES de REGIO MONTE dictus
alias MÜLLERVS.
Gnigmis Mathematicus et de
Re Typographica Nörnbergensium
Optime meritus.
Nat: A 1436 d. 6 Juny   Den: A. 1476 d. 6 July Aet: XLI.
Ex collectione Friderici Roth – Scholtzÿ Norimberg.

The resulting book was published some years later as an *Epitome of Ptolemy's Almagest,* and it was of enormous importance. Regiomontanus had revealed so many errors in the Alfonsine Tables that he exclaimed, "The common astronomers of our age are like credulous women, receiving as something divine and immutable whatever they find in books, for they make no effort themselves to find the truth." The motions of all the planets were found to be incorrectly plotted in the tables, so that their appearances, conjunctions, eclipses, and so on all might take place many degrees away from the place predicted and days before or after the time predicted.

BELIEF WITHOUT QUESTION

The interesting thing about Regiomontanus's work, however, is that he did not question the mathematical models that he found in Ptolemy. He assumed they were correct, while the observations of positions underlying the Alfonsine Tables were wrong.

The reform of astronomy, Regiomontanus argued, must come from far more accurate observations from which the paths of the planets could be worked out precisely, still using Ptolemy's theories. Regiomontanus set in train a process of reform in astronomy whose end would have astounded him: He revealed discrepancies between observed reality and the standard texts that would only be explained after the revolution wrought by Copernicus, Tycho Brahe, and Kepler. Copernicus himself studied Regiomontanus's work and was struck by the problems it revealed, but he concluded that it was the Ptolemaic framework itself that was at fault.

Regiomontanus settled in Nuremberg and established a printing press with which he intended to disseminate scientific works; but he, too, died prematurely. He was a transitional figure who stood on the threshold of a new science. Although he did not dream of questioning Ptolemy's cosmic model, he introduced an important new note of empiricism into astronomy, looking at the heavens with fresh eyes.

Above: An imaginary portrait of Ptolemy with his astrolabe, from the *Nuremberg Chronicle.*

Opposite: A Renaissance astronomer measuring the altitude of a star with a quadrant.

Renaissance Astronomy:
Copernicus and His Revolution

Nicholas Copernicus, whose ideas changed peoples' understanding of the universe.

In the year 1543—many decades after the invention of printing, after the discovery of the New World, and after the rebirth of art—there occurred an intellectual event that changed forever humans' understanding of themselves and their world: Copernicus published his theory that the Earth was not the center of the universe, but was a planet orbiting the Sun.

The idea that the Earth was the central point of the cosmos had seemed to follow from the evidence of our own eyes, and it had been universally accepted by all thinking people. The rejection of that idea was one of the great landmarks in humankind's intellectual history, whose profound implications required many years before they were fully understood.

Nicholas Copernicus (1473–1543) was a mathematical astronomer born in the town of Torun, which stood then on the borders between Germany and Poland. His original name was Nikolaj Kopernik, but he Latinized his name, as did many scholars at this time. Copernicus was a churchman by profession and an astronomer only by inclination, and his role in the revolution that bears his name is not as simple as is sometimes supposed.

ELEGANT THEORY

The natural model of a scientific revolution is to imagine scientists working within an agreed framework. They are confronted with new data that proves impossible to reconcile with that framework, so they are forced into a new understanding of their subject. Yet nothing like this happened in the case of Copernicus. New facts, new astronomical observations, new evidence—all these are absent from his work. He was a student of books rather than of nature, and observation of the sky was not the basis of his new theory. Instead, he performed a highly original thought experiment: He devised a new geometric model that accounted for the movements that we see in the heavens in a simpler, more elegant way than the classical theory of Ptolemy had.

Copernicus did not leave a detailed account of the way his thinking developed, but he said enough for us to know that he was dissatisfied with the Ptolemaic theory of the heavens from an early stage in his career. Its complex system of epicycles and eccentrics (see Volume 1, page 52) seemed to Copernicus to be impossible to reconcile with the movements of real physical objects in space. These were doubts that many astronomers had experienced over the

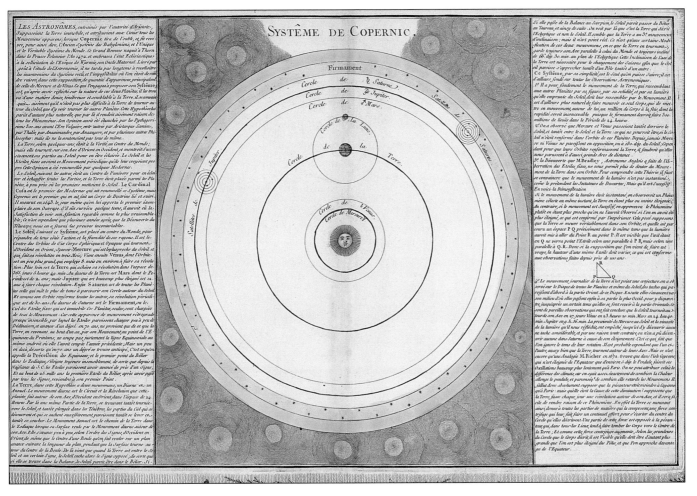

The Copernican system with the Sun at the center of the cosmos.

centuries, but none of them had ever seen the correct way out. It occurred to Copernicus, however, that the movement may be real or apparent: We can easily see how apparent movement may be caused by the observer's own movement—for example, if we sit in a spinning chair, the room appears to revolve around us. This gave Copernicus his vital clue, and he later wrote:

"A seeming change of place may arise from the motion of either the object or the observer . . . . If then some motion of the Earth be possible, the same will be reflected in external bodies, which must seem to move in the opposite sense . . . . I began to think of the mobility of the Earth; and though the idea seemed absurd . . . . I considered that I might find, by assuming some motion in the Earth, sounder explanations for the revolutions of the celestial spheres."

And so it proved, for looking first at the movements of Mercury and Venus, Copernicus saw that their constant proximity to the Sun was much more easily explained by supposing that they revolved around the Sun and nearer to it than the Earth. But Copernicus still accepted the physical reality of the celestial spheres: Therefore it was impossible that these two planets should orbit the Sun while the Sun and the other still orbited the Earth, for the spheres would have to intersect. Only by placing the Sun at the center of the system and

61

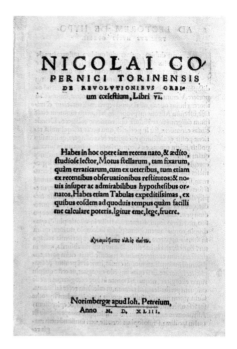

The title page of Copernicus's great work *On the Revolutions of the Heavenly Spheres*, printed in Nuremberg in 1543.

NICOLAUS COPERNICUS (1473–1543)

- Originally named Nikolaj Kopernik.
- Founder of modern astronomy.
- Born Torun, Prussia (now Poland).
- 1491–94 Studied mathematics and optics at Cracow University.
- 1496 Went to Italy to study canon law and attended astronomy lectures at Bologna University.
- 1497 Nominated a canon at Frombork cathedral—but never took holy orders.
- 1501–05 Learned medicine at Padua. While there, in 1503, made a doctor of canon law by the University of Ferrara.
- Returned to Frombork, Poland, as an administrator and the bishop of Ermeland's (who was also his uncle) medical advisor.
- Deeply troubled by Ptolemy's assertion that the Earth was the center of the universe.
- 1512 Started a mathematical explanation of the Sun being the center of the universe, but unwilling to publish for fear of the church establishment.
- 1543 By now old and ill, he was reluctantly persuaded by his pupil Rheticus to finally publish his theories in *De Revolutionibus Orbium Coelestium* ("On the Revolutions of the Heavenly Spheres"). He dedicated the work to Pope Paul III. Copernicus received his copy on his deathbed.
- The book scandalized the establishment and was immediately banned by the Catholic church and placed on the list of forbidden books, where it remained until 1835.

the Earth as a planet in motion around it could all movements of the planets as we see them be explained.

Copernicus still had to account for certain irregularities in the planets' paths—which we know are caused by the fact that their orbits are not perfectly circular. Copernicus did this in the traditional way by inventing more epicycles (see Volume 1, page 55), showing that he was still approaching the problem in a purely mathematical way. Galileo would later point out that the path ascribed by Copernicus to Venus could not explain that planet's constant brilliance, which showed that personal observation of the sky played no part in Copernicus's method. The Copernican revolution was a purely *conceptual* one. It was not a *discovery,* for he was never able to offer any evidence that it was true. Other mathematicians could only consider his model and try to assess how it accounted for what we see in the sky. And in any such assessment the idea that the Earth was spinning through space seemed to contradict our common experience, which tells us that the Earth is at rest.

RELUCTANCE TO PUBLISH

Copernicus's theory was essentially complete by the year 1510, and within a year or two of that date he had circulated it in manuscript among a few friends. Long years passed, during which Copernicus must have pondered the implications of his ideas, but he did not publish them. There is one very revealing detail about his thought processes during these years. In his original manuscript, which still survives, Copernicus referred to the ancient Greek astronomer Aristarchus, who had suggested the same idea of a Sun-centered universe long before. But Aristarchus had narrowly escaped being put on trial for his outrageous and irreligious views, and Copernicus canceled this reference

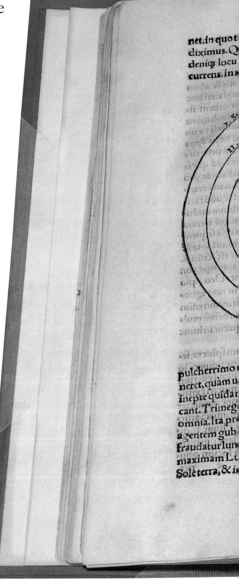

when his book was finally published. There seems little doubt that he hesitated to publicize his idea because he was aware how controversial it would be, and in particular he must have foreseen trouble with the church authorities.

We cannot be certain that he would ever have published it at all had not his hand been forced by one of his friends, George Rheticus, who in 1540 printed his own brief account of the new theory. Copernicus naturally wanted to give his theory in full in his own way, so he set to work to prepare the long-withheld manuscript for the press. The book was printed in Nuremberg in 1543, with the title *De revolutionibus orbium coelestium* ("On the Revolutions of the Heavenly Spheres"). Whether the famous story is true that a copy was placed in Copernicus's hands on May 25 of that year, as he lay dying, we can never be sure. We only know that he died in almost complete obscurity, presumably unaware that his name would be linked with one of the greatest intellectual revolutions in history.

Copernicus's drawing of the Sun-centered cosmos illustrates his *De revolutionibus orbium coelestium.*

Renaissance Astronomy:
The Response to Copernicus

The Copernican revolution was one of the great turning points in intellectual history, but it is important to understand that it was not an event like the French Revolution, which changed the world in a few days. Copernicus's new theory of the solar system appeared in print in the early summer of 1543, but more than half a century was to pass before it was widely accepted by scientists and scholars, and even longer before its implications were fully worked out.

Why was this? The first objection to Copernicanism was the most obvious one, that it entailed belief in an Earth that was whirling through space and rotating on its axis. How could this be true, contemporaries asked, when their senses told them that the Earth was motionless? If an object such as a stone is thrown up into the air, it does not land miles away as the surface of the Earth rushes by beneath it.

As one philosopher wrote: "No one in his senses will ever think that the Earth, heavy and unwieldy from its own weight and mass, staggers up and down around its own center or that of the Sun, for at the slightest jar of the Earth, we would see cities and fortresses, towns and mountains thrown down."

Only gradually did it come to be understood that this does not happen because everything on the Earth is *sharing* the Earth's motion. The stone thrown vertically up into the air is moving with the Earth and therefore returns to the same point that it left. The stone has moved in space in other ways, too, but this would not be apparent to any observer standing on the spot who also shares the Earth's motion. It was Galileo who made this clear much later through his analogy of an object dropped from the mast of a moving ship: Everyone would expect the object to land at the foot of the mast, not some yards away as the ship moves. So this objection, although at first so powerful, was capable of a straightforward solution.

INFINITE UNIVERSE

The second problem was more subtle and more profound, and it concerned the scale of the universe. The classical universe of Ptolemy was very large (see Volume 2, page 52), but it was *finite*; it was closed by the outermost sphere, which carried the stars. The stars were said to be fixed because their positions on that sphere never changed in relation to each other. But if the Earth was indeed moving in an orbit around the Sun, then the Earth's position would

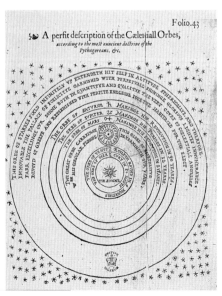

The Copernican system presented in English by Thomas Digges in 1576. Digges explained the crucial ideas that the sphere of the stars "extended infinitely up" and that the stars were actually suns.

change by many of millions of miles in the course of the year. One of the effects of this movement should be that the positions of the stars should change enormously as viewed from the Earth: This would be an effect of <u>parallax</u>. But Copernicus noticed, and others after him noticed too, that no such parallax effect occurred.

Even Tycho Brahe, the most dedicated astronomical observer of his age, searched in vain for any such parallax changes in stellar positions. The only possible explanation for this was that the stars were so far away that no parallax shift could be detected. This, in turn, implied that the stars were far, far further from the Earth than had ever been imagined, that the scale of the universe was much vaster than had been supposed. Moreover, there was no longer any reason to believe that the Earth or the Sun lay at the center of the universe. Therefore the human world was displaced from being the focus of the whole creation to being a random point in a huge and uncharted universe.

"The heavens," concluded Copernicus, "are immense and present the appearance of an infinite magnitude." The crucial word here is "infinite." The idea of an infinite universe was to have an immense

Copernicus's revolutionary ideas raised new questions about the stars: What were they, where were they, how was man to understand his place in this new universe? This is the wonderful ceiling fresco of the constellations in Caprarola, Italy.

A celestial globe, made in Nuremberg by Johannes Schoner in 1533 and said to be one of the oldest to be found.

impact, especially on religious thought. Why should God create an infinite universe? If the great goal of history was the salvation of humankind, why should that take place in an infinite space?

The idea that the Earth is in motion and the idea of an infinite universe both directly contradicted classical Aristotelian science, and that was the real reason for resistance to Copernicanism. All the scholars and philosophers in Europe had been trained to accept the picture of the universe as a finite system enclosed within a nest of spheres, a mechanism ruled by God. This was the universe of the medieval church, of St. Thomas Aquinas and Dante, and it was comprehensible and rational, while the new universe of Copernicanism was neither.

Aristotle had taught that falling objects were seeking the center of the universe, which was also the center of the Earth, and that all natural motion is to be explained in this way except the spherical motion of the heavens, where other laws exist. But if the Earth were not the center of the universe, why should heavy objects fall downward at all? In the Aristotelian universe the heavier elements—earth and water—gather at the center, while the lighter—air and fire—rise upward. But again, if the Earth is not the center, this could not be true. Copernicus suggested instead a principle of cohesion in which all heavy matter would gather together into a sphere, although this need not be the center of the universe. But if the Earth is a planet, might not all the other planets be earths? Might not the Aristotelian division of the universe into two realms, the Earth and the heavens, be a fable?

Behind all these questions there loomed the ultimate problem. If the Earth is not the still center of a finite universe, what power keeps the vast mass of the Earth and all the other planets in motion around the Sun? Copernicus's theory was still couched in geometric terms, it was abstract and conceptual, but it implied the need for a new kind of physics. The Christian church had adopted the Aristotelian view of the universe and had built it into orthodox religious thought. When that view was challenged, as it now was, the challenge might appear to be directed against the entire structure of Christian thought. The implications of Copernicanism took many years to sink in and be fully understood, but this shift in humans' vision of the universe had a profound effect not only on science but on religion and philosophy.

Science in the Renaissance:
A Retrospect

The period that we have been examining in this volume stands at the threshold of modern history. The century of the Renaissance from 1450 to 1550 saw the European understanding of the world change out of all recognition. Cities were rebuilt; the unity of Christendom was broken up; Europeans explored the entire world and discovered a new continent; communication was revolutionized by the invention of printing. In comparison with what had gone before, this was an age of change and of dynamism. Yet we cannot tie all these changes together or say that they all arose from a single cause or from some new intellectual force. Strangest of all perhaps, there was no equivalent revolution in pure science, in the way that humans understood the workings of nature. The Copernican revolution began only at the very end of this period, and its effects took many years to be felt.

Humans looked at nature with fresh eyes, but this change was seen in certain fields of applied science: in drawing, in navigation, in mapmaking, in metallurgy, in botany and zoology, and in anatomy. Yet there was no conceptual revolution underlying the new approach to all these fields. If there was one intellectual current that was stronger than any other, it was the resurgence of the magical philosophy of nature under the influence of Hermeticism. This philosophy was like a science to the extent that it taught that nature was governed by hidden laws; but the difference was that these laws were secret—they had been passed down from ancient sources, and they were to be learned from books, not by reason or experiment. Moreover, these laws worked because behind the face of nature lay a host of personalized forces, spirits, and demons that could be commanded by the expert in magic.

What was missing from the intellectual climate was the philosophy of empiricism, the principle that ideas and theories should be tested by experience and experiment. The thinking of the sixteenth century was still dominated by authority, by the orthodoxies of the past. The problem of the Renaissance was the

Renaissance artists looked at the richness of nature with fresh eyes. This is a salamander from the *Margarita Philosophica* ("The Philosophic Pearl") of 1503.

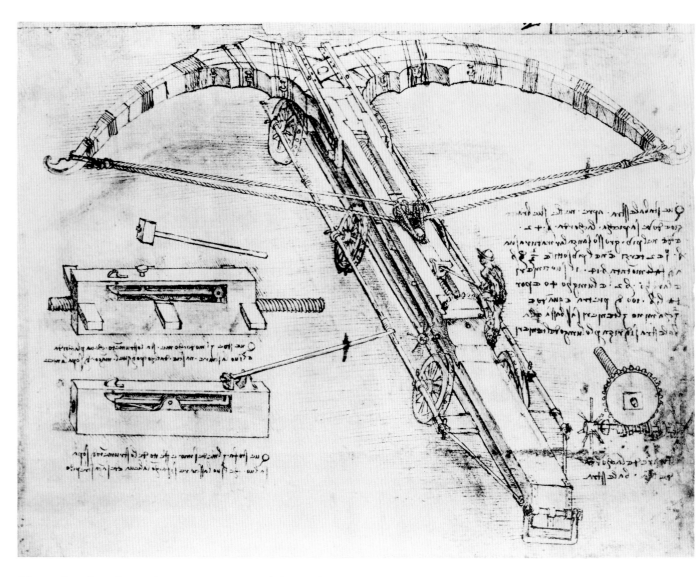

Above: A colossal crossbow designed by Leonardo.

Opposite, Above and Below: The scientific and intellectual revolution of the late Renaissance had its roots in astronomy. The top illustration shows an astronomer at work in the mid-sixteenth century. Note the armillary at the extreme right of the frame. The lower picture is entitled "The Comet of 1532." It would not be until the seventeenth century and the work of Halley (see Volume 5, page 24) that comets would be explained. Even Galileo thought that they were optical illusions.

problem of a mismatch between society and learning. Society was dynamic, while learning was static. While explorers and inventors were changing humans' perception of the world around them, the universities of Europe were teaching logic, mathematics, and theology just as they had for centuries. Here the universe was seen through Aristotle's eyes, the heavens through those of Ptolemy, the human body through Galen's. This can easily be seen if we consider the world view of Shakespeare and Marlowe, who in the 1590s could assume that their audience believed in the heavenly spheres, hell, and purgatory, the four elements, witchcraft, and divination, the music of the spheres, astrology, and alchemy; there is no evidence that either man had ever heard of Copernicus. Mystics and magicians such as Bruno, Agrippa, and Campanella are important because they tried to break out of the sterile world of medieval learning into a more personal understanding of nature's powers. They are an integral part of the Renaissance rediscovery of nature, but that rediscovery had not yet found a genuinely empirical language of measurement and experiment that could begin to bring order to the diversity of nature's forms and processes.

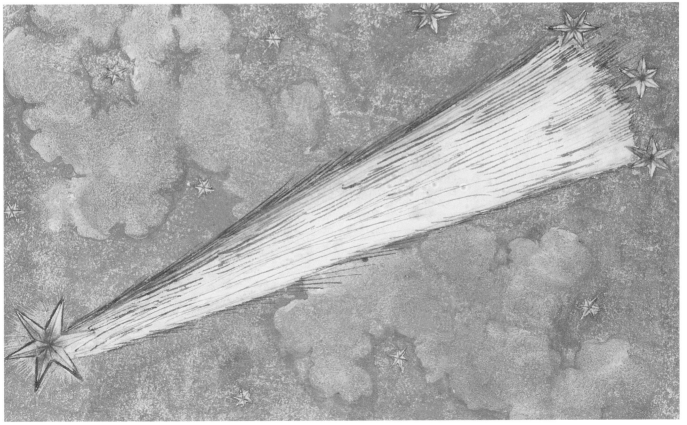

Glossary

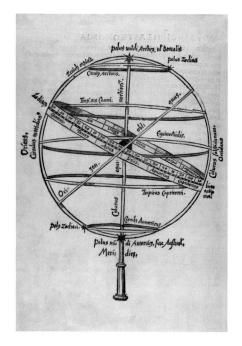

Armillary sphere from a handbook of 1535.

armillary an old astronomical instrument made of rings showing the positions of important circles in the celestial sphere.

algebra a form of mathematics in which letters representing numbers are combined according to the rules of arithmetic.

algorithm step-by-step procedure for solving a problem.

almanac a publication containing astronomical data for a given year.

antitrinitarian having no belief in the unity of Father, Son, and Holy Spirit as three persons in one God.

astrology the divination of the influence of the stars on humans.

astronomy the study of what lies outside the Earth's atmosphere.

atom one of the minute particles from which the universe is made.

azimuth usually in astronomy and navigation an arc of the horizon measured between a fixed point and the vertical circle passing through the center of an object clockwise through the north point 360°; horizontal direction expressed as the angular distance between the direction of a fixed point and the direction of an object.

chemistry science that deals with the composition, structure, and properties of substances and the transformations that they undergo.

chronology science dealing with measuring time by regular divisions and giving events their proper dates.

cosmos literally an ordered structure; a harmonious, systematic universe

cyclical moving in cycles; relating to cycles.

doctrine a teaching.

dynamics branch of mechanics that deals with forces and their relation to the motion of bodies, sometimes also to the equilibrium of bodies.

empirical based on observation, experience, or experiment.

equinox either of the two times each year when the Sun crosses the equator, and the days and nights are everywhere of equal length.

geometry branch of mathematics dealing with the measurement and properties of points, lines, angles, surfaces, and solids.

guild a medieval association of craftspeople.

Captions to illustrations on page 72.

Above: Plate from a mid-sixteenth-century Portuguese handbook of navigation showing phases of the Moon.

Below: Title page of the *Margarita Philosophica* ("Philosophic Pearl").

hydrostatic relating to fluids at rest and the pressures they exert.

hygrometer an instrument for measuring the humidity of the atmosphere.

incommensurability given a ruler with equally spaced markings, all three sides of a triangle cannot be measured exactly.

impetus the property possessed by a moving body by virtue of its mass and its motion.

Inquisition a medieval Roman Catholic tribunal for discovering heretics.

irrational number a number that can be expressed as an infinite decimal with no set of consecutive digits repeating itself indefinitely and that cannot be expressed as the quotient of two integers.

kinematics the description of motion.

logarithm the exponent that indicates the power to which a number is raised to produce a given number.

magnetism a science that deals with magnetic phenomena.
mechanics a branch of physics that deals with energy and forces and their effect on bodies.
metamorphic having pronounced change through the application of pressure, heat, or water.

nadir the point of the celestial sphere opposite the zenith and vertically downward from the observer.

occult involving supernatural powers.
optics a science that deals with light.

parallax the difference in apparent direction of an object as seen from two different points not on a straight line; the angular difference of direction of a heavenly body measured from two points on Earth's orbit.
projection intersecting coordinate lines on a flat surface on which the curved surface of a sphere may be mapped.

resistance an opposing or retarding force; opposition of a body to an electric current passing through it.

specific gravity the ratio of the density of a substance to the density of another (such as water).
spherical trigonometry trigonometry applied to spherical triangles and polygons.
statics mechanics dealing with forces that produce equilibrium between two bodies.

theology the study of religious faith.
theorem a rule that provides proof of a theory.
theory an idea that has still to be proved.
tolerance the allowable variation from a standard.
transcendent beyond the limits of ordinary experience.
triangulation determination of diagonal distances by means of bearings from two fixed points a known distance apart.
trigonometry the study of the property of triangles.

velocity the rate of change of position along a straight line in relation to time.
vortex circular motion that tends to form a cavity or vacuum at the center.

zenith the point of the celestial sphere that is directly opposite the nadir and vertically above the observer.

Captions to illustrations on page 73.

Top: A crab from Gesner's *Historia Animalium* ("History of Animals"), 1558; Gesner gave faithful pictures of thousands of species alongside the mythical creatures of the past.

Center: A model of one of Leonardo's pencil drawings: This is a lens grinder in the Science Museum, London.

Bottom: Title page of Regimontanus's *Almagest*.

Vesalius's anatomy of the eye from *De humani corporis fabrica* ("The Structure of the Human Body") of 1543.

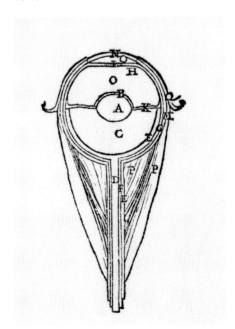

PERIOD	1450	1460	1470	1480	1490	1500	1510

WORLD EVENTS

1453 Turks capture Constantinople (Istanbul). End of the Byzantine Empire.

1478 Czar Ivan III frees Russia from the Mongols.

1488 Portuguese sail into the Indian Ocean.

1492 Columbus reaches America.

1492 Fall of Granada to the Spanish. End of Moslem power in Spain.

1493 Treaty of Tordesillas divides New World between Spain and Portugal.

1497 Cabot discovers Newfoundland.

1498 Portuguese land in India.

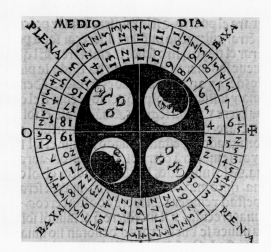

SCIENCE

1455 Gutenberg prints first books in Mainz.

1472 Valturio's *De re militari* ("On Military Matters").

1477 First world atlas published in Rome.

1480–1510 da Vinci active.

1491–93 Ketham's *Fasciculus de Medicinae* Medical Papers"), first medical textbook, published.

1496 Regiomontanus' printed version of Ptolemy's *Almagest* published.

1503 Reisch's *Margarita Philosophica* ("PhilosophicPearl") published.

1509 Pacioli's *On Divine Proportion* published. The text is illustrated by Leonardo.

1510–30 Paracelsus—medicine and alchemy.

ART & CULTURAL EVENTS

1440–90 Medici family rules Florence.

1460 on Printing spreads to Italy, Germany, France, and Switzerland.

1470s Botticelli, artist, active.

1490–1520 Erasmus, humanist scholar active.

1490–1520 Dürer active.

1493 Nuremberg Chronicle printed.

1495 Leonardo's *Last Supper* painted.

1507 Waldseemuller's world ma names America after Amerig Verspucci.

1508–12 Michelangelo's Sistin Chapel painted in Rome.

1513 Machiavelli publishes *The Prince*.

1516 Thomas More's *Utop*

1520	1530	1540	1550	1560	1570	1580	1590	1600

1519–22 Magellan's first circumnavigation of the Earth.

1519 Luther begins the Reformation.

1519–20 Cortes conquers the Aztec Empire.

1520–40 Turks conquer eastern Europe.

1532 Pizarro conquers the Inca Empire.

1534 Cartier discovers Canada.

1534 Henry VIII separates English church from Rome.

1540 Jesuit order founded.

1545 Council of Trent—start of Counter-Reformation.

1550–80 Ivan the Terrible rules Russia.

1560–1600 Elizabethan age in England.

1560–90 French wars of religion.

1571 Battle of Lepanto—end of Turkish power in the Med.

1571 Spanish conquer Philippines.

1572 Dutch revolt against Spain.

1585–90 Failure of English colonies in Virginia.

1600 British and Dutch found companies to trade in the East Indies.

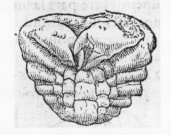

1530 Brunfels's illustrated botany published.

1533 Agrippa's *Occult Philosophy* published.

1540 Gersdorff's *Field Book of Wounds* published.

1540 Jesuit order founded.

1542 Fuch's *History of Plants* published.

1543 Copernicus's *The Revolutions of the Heavenly Spheres* published.

1543 Vesalius's *The Fabric of the Human Body* published.

1544 Münster's *Cosmographia* published.

1545 Medina's *Art of Navigation* published.

1551–56 Gesner's *Zoology* published.

1552 Cardano publishes horoscope of Christ.

1553 Michael Servetus burned for heresy.

1556 Agricola's *De re metallica* ("On Metallurgy") published.

1569 Mercator's world map with mathematical projection for navigators.

1571 Digges's applied math textbook *Pantometrica* published.

1572 Pare's *Treatise on Surgery* published.

1572 Tycho observes supernova and starts astronomical career.

1590s Galileo works on physics.

1600 Bruno burned for heresy.

1600 Campanello in prison working on *The City of the Sun*.

1520s Hampton Court palace built outside London, England.

1520s Fontainebleau palace built outside Paris, France.

1545–65 St. Peter's built in Rome, Italy.

1560–90 Montaigne, French philosopher.

1560–80 Palladio's neoclassical architecture in Italy.

1565–85 Escorial palace, Madrid, Spain built.

1570–1600 Painter El Greco active.

1590s Shakespeare's early plays.

1590s Francis Bacon, first philosopher of science, active.

73

Resources

FURTHER READING

There is a wealth of books published on the history of science, particularly biographies of great scientists. The following list includes many large works that contain many further resources.

Adams, F.D.: *The Birth and Development of the Geological Sciences*; Dover Publications, 1955.

Bowler, P.: *Evolution: The History of an Idea*; University of California Press, 1998.

Bowler, P.: *The Norton History of Environmental Sciences*; W.W. Norton & Co., 1993.

Boyer, C. & Merzbach, U.: *A History of Mathematics*; John Wiley & Sons Inc., 1989

Brock, W.H.: *The Fontana History of Chemistry*; Fontana Press, 1992.

Butterfield, H.: *The Origins of Modern Science*; Free Press, 1997.

Clagett, M.: *Greek Science in Antiquity*; Dover Publications, 2002.

Cohen, I.B.: *Album of Science: From Leonardo to Lavoisier*; Charles Scribner's Sons, 1980.

Cohen, I.B.: *The Birth of a New Physics*; W.W. Norton, 1985.

Crombie, A.C.: *Augustine to Galileo: The History of Science AD400–1650*; Dover Publications, 1996.

Crombie, A.C.: *Science, Art and Nature in Medieval and Modern Thought*; Hambledon, 1996.

Crosland, M.: *Historical Studies in the Language of Chemistry*; Heinemann Educational, 1962.

Eves, H.: *An Introduction to the History of Mathematics*; Thomson Learning, 1990.

Gillispie, C.C. (ed.): *Concise Dictionary of Scientific Biography*; Charles Scribner's Sons, 2000.

Gillispie, C.C.: *Genesis and Geology*; Harvard University Press, 1996.

Hallam, A.: *Great Geological Controversies*; Oxford University Press, 1983.

Ihde, A.J.: *The Development of Modern Chemistry*; Dover Publications, 1983.

Jaffé, B.: *Crucibles: The Story of Chemistry from Alchemy to Nuclear Fission*; Dover Publications, 1977.

Jungnickel, C. & McCormmach, R.: *Intellectual Mastery of Nature: Theoretical Physics from Ohm to Einstein*; University of Chicago Press, 1986.

Koyré, A.: *From the Closed World to the Infinite Universe*; The Johns Hopkins University Press, 1994.

Kuhn, T.: *The Copernican Revolution: Planetary Astronomy in the Development of Western Thought*; Harvard University Press, 1957.

Lindberg, D.C.: *The Beginnings of Western Science*; University of Chicago Press, 1992.

Porter, R. (Ed.): *The Cambridge Illustrated History of Medicine*; Cambridge University Press, 1996.

McKenzie, A.E.E.: *The Major Achievements of Science*; Iowa State Press, 1988.

Morton, A.G.: *A History of Botanical Science*; Academic Press, 1981.

Nasr, S.H.: *Islamic Science—An Illustrated Study*; London, 1976.

North, J.D.: *The Fontana History of Astronomy and Cosmology*; Fontana Press, 1992.

Olby, R. (et al.): *A Companion to the History of Modern Science*; Routledge, 1996.

Parry, M. (ed.): *Chambers Biographical Dictionary*; Chambers Harrap, 1997.

Porter, R.: *The Greatest Benefit to Mankind: a Medicinal History of Humanity from Antiquity to the Present*; HarperCollins, 1997.

Roberts, G.: *The Mirror of Alchemy*; British Library Publishing, 1995.

Ronan, C.A.: *The Cambridge Illustrated History of the World's Science*; Cambridge University Press, 1983.

Ronan, C.A.: *The Shorter Science and Civilisation in China*; Cambridge University Press, 1980.

Selin, H. (ed.): *Encyclopedia of the History of Science, Technology and Medicine in Non-Western Cultures*; Kluwer Academic Publishers, 1997.

Uglow, J.: *The Lunar Men*; Faber and Faber, 2002.

Van Helden, A.: *Measuring the Universe: Cosmic Dimensions from Aristarchus to Halley*; University of Chicago Press, 1985.

Walker, C.B.F. (ed.): *Astronomy before the Telescope*; British Museum Publications., 1997.

Whitfield, P.: *Landmarks in Western Science: From Prehistory to the Atomic Age*; The British Library, London, 1999.

Whitney, C.: *The Discovery of Our Galaxy*; Iowa State University Press, 1988.

THE INTERNET

Websites relating to the history of science break down into four types:

- Museum sites that offer some history and artifact photography. This is often the easiest way to visit international sites or those of states too far away to get to in person.
- College or other educational establishment sites that often provide online learning or study resources.
- General educational sites set up by enthusiasts (often teachers) and historians.
- Societies or clubs.

Examples of these three types of website include:

Museums

http://www.mhs.ox.ac.uk/
Museum of the History of Science, Oxford, England.
Housed in the world's oldest surviving purpose-built museum building, the Old Ashmolean.

http://www.mos.org/
Museum of Science, Boston.

http://www.msichicago.org/
Museum of Science and Industry, Chicago.

http://www.lanl.gov/museum
Bradbury Science Museum, a component of Los Alamos National Laboratory.

http://www.si.edu/history_and_culture/history_of_science_and_technology/
Smithsonian Institution site.

http://www.sciencemuseum.org.uk/
National Museum of Science and Industry, London, England.

http://galileo.imss.firenze.it/
Institute and Museum of the History of Science, Florence, Italy.

http://www.jsf.or.jp/index_e.html
Science Museum, Tokyo.

Colleges or institutions

http://sln.fi.edu/tfi/welcome.html
Franklin Institute with online learning resources and study units.

http://www.fas.harvard.edu/~hsdept/
Department of the History of Science of Harvard University.

http://www.hopkinsmedicine.org/graduateprograms/history_of_science/
Department of the History of Science, Medicine and Technology at Hopkins.

http://www.lib.lsu.edu/sci/chem/internet/history.html
Louisiana State University provides excellent history of science internet resources and links.

http://dibinst.mit.edu/
The Dibner Institute is an international center for advanced research in the history of science and technology and located on the campus of MIT.

http://www.mpiwg-berlin.mpg.de/ENGLHOME.HTM
Max Planck Institute for the History of Science

http://www.princeton.edu/~hos/
History of Science @ Princeton.

http://www.astro.uni-bonn.de/~pbrosche/hist_sci/hs_sciences.html
History of sciences from Bonn University, Germany, including indexes on the history of astronomy, chemistry, computing, geosciences, mathematics, physics, technology.

Educational sites

http://echo.gmu.edu/center/
ECHO—Exploring and Collecting History Online—provides a centralized guide for those looking for websites on the history of science and technology.

http://www.wsulibs.wsu.edu/hist-of-science/bib.html
Provides reference sources in the form of bibliographies and indexes.

http://dmoz.org/Society/History/By_Topic/Science/Engineering_and_Technology/
Open Directory Project providing bibliography and links.

http://orb.rhodes.edu/
ORB—the Online Reference Book—provides textbook sources for medieval studies on the web. It includes the Medieval Technology Pages—providing information on technological innovation and related subjects in western Europe—and Medieval Science Pages, a comprehensive page of links to medieval science and technology websites.

http://www.fordham.edu/halsall/science/sciencesbook.html
This page provides access to three major online resources, the Internet Ancient History, Medieval, and Modern History Sourcebooks.

http://www2.lib.udel.edu/subj/hsci/internet.htm
The University of Delaware Library provides an excellent guide to Internet resources.

Societies

www.hssonline.org
History of Science Society provides for its members the History of Science, Technology, and Medicine Database—an international bibliography for the history of science, technology, and medicine.

http://www.chstm.man.ac.uk/bshs/
British Society for the History of Science.

Set Index